Math Worlds

SUNY Series in Science, Technology, and Society
Sal Restivo, Editor

Math Worlds

Philosophical and Social Studies of Mathematics and Mathematics Education

Edited by
Sal Restivo
Jean Paul Van Bendegem
and
Roland Fischer

STATE UNIVERSITY OF NEW YORK PRESS

Published by
State University of New York Press, Albany

© 1993 State University of New York

For information, address State University of New York
Press, State University Plaza, Albany, N.Y., 12246

Production by E. Moore
Marketing by Dana E. Yanulavich

Library of Congress Cataloging-in-Publication Data

Math worlds : philosophical and social studies of mathematics and
    mathematics education / Sal Restivo, Jean Paul van Bendegem, and
    Roland Fischer.
        p.   cm. — (SUNY series in science, technology, and society)
    Includes index.
    ISBN 0–7914–1329–2 (alk. paper). — ISBN 0–7914–1330–6 (pbk. :
    alk. paper)
        1. Mathematics—Philosophy. 2. Mathematics—Social aspects.
    3. Mathematics—Study and teaching.   I. Restivo, Sal P.
    II. Bendegem, Jean Paul van, 1953–   . III. Fischer, Roland, 1945–   .
    IV. Series.
    QA8.6.M376   1993
    510' .1—dc20

10 9 8 7 6 5 4 3 2 1

Contents

PART I: GENERAL INTRODUCTION

1

The Promethean Task of
Bringing Mathematics to Earth

PROLOGUE

Whenever someone wants to give an example of a truth that is absolutely certain and indubitable, he or she is likely to use the Pythagorean theorem or a simple equation like 2 + 2 = 4. Martin Gardner (1981), for example, challenged the efforts of mathematicians such as Davis and Hersh (1981) and Kline (1980) to "undermine" the certainty of mathematics by giving the following example. In prehistoric times, "2 + 2 = 4" was "accurately modeled" whenever two dinosaurs met two dinosaurs in spite of the facts that there were no humans to observe the event and that the dinosaurs were incapable of comprehending or representing their gathering mathematically. This strategy appears across the entire range of cultural thought. The novelist Thomas Hardy wrote these words for Jude in *Jude the Obscure*:

> Is a woman a thinking unit at all, or a fraction always wanting its integer? How you argued that marriage was only a clumsy contract—which it is—how you showed all the objections to it—all the absurdities! If two and two made four when we were happy together, surely they make four now? I can't understand it, I repeat! (1969:370)

And the social theorist Karl Mannheim helped to keep a whole gener-

3

ation of sociologists of science outside the inner sanctum of "objectivity" when he wrote:

> Even a god could not formulate a proposition on historical subjects like $2 \times 2 = 4$, for what is intelligible in history can be formulated only with reference to problems and conceptual constructions which themselves arise in the flux of historical experience. (1936:79)

The mathematician, historian, and Marxist scholar Dirk Struik, a founder of the sociology of mathematics, referred to *both* the Pythagorean theorem *and* $2 \times 2 = 4$ in this defense of realism:

> Our conviction of the eternal validity of Pythagoras' theorem, of the fact that $2 \times 2 = 4$, is not based on some *a priori* conception, nor can it be shaken by any clever mathematician who in a big book with formulas concludes that these theorems are mere conventions. Our conviction is based on the fact that the theorems correspond to properties of the real world outside our consciousness which can be tested, and are accessible for testing to all persons from their earliest childhood. (1949:146–47)

But realism has not kept the nemesis of $2 + 2 = 4$, $2 + 2 = 5$, at bay. During the era of five-year plans in the Soviet Union, $2 + 2 = 5$ appeared. It was not designed as a serious threat to realism but rather to express the hope that the five-year goals might be achieved in four years. It is more interesting to see how the novelists Orwell and Dostoevsky used $2 + 2 = 4$ and $2 + 2 = 5$ to represent social and political systems, but in opposite ways. In Orwell's *1984*, O'Brien tells Winston that two and two are four *sometimes*: "Sometimes, Winston. Sometimes they are 5; sometimes they are 3; sometimes they are all of them at once. You must try harder. It is not easy to become sane" (1956:201). For Orwell, $2 + 2 = 4$ is a certainty against which to measure the totalitarian extremes of Big Brother, who is represented by $2 + 2 = 5$. Dostoevsky, on the other hand, uses $2 + 2 = 5$ in *Notes From Underground* to represent a challenge to rigid and routinized social and political realities:

> . . . twice-two-makes-four is not life, gentlemen. It is the beginning of death. Twice-two-makes-four is, in my humble opinion, nothing but a piece of impudence . . . a farcical, dressed up fellow who stands across your path with arms akimbo, and spits at

you. Mind you, I quite agree that twice-two-makes-four is a most excellent thing: but if we are to give everything its due, then twice-two-makes-five is sometimes a most charming little thing too. (n.d.:139)

There is a noteworthy coincidence between this passage and Oswald Spengler's (1926:55–58) notion of number as an exemplar of the "become," the "hard set," and "Death." Spengler, a mathematics teacher, also offered a challenge to those who accepted the self-evidence of certain number facts:

Even the most "self-evident" propositions of elementary arithmetic such as 2 × 2 = 4 become, when considered analytically, problems, and the solution of these problems was only made possible by deductions from the theory of aggregates, and is in many points still unaccomplished. (1926:84)

If there are readers who think these sorts of oppositions can only take place outside of mathematics proper, let them consider the following examples. Jourdain, for example, claimed that "Somebody might *think* that 2 and 2 are 5: we know by a process which rests on the laws of Logic [which refer to 'Truth'], that they make 4" But he almost immediately indicates that things may be more complicated. He claims that 1+1=2 may be "mistakenly written." This notation makes it look as if there are *two* whole classes of unit classes. In fact, there is only one, 1. 1 is a class of certain classes. Therefore, 1 + 1 = 2 means that "if x and y are members of 1, and x differs from y, then x and y together make up a member of 2." (Jourdain, 1956:67–71).

Now consider that Bertrand Russell viewed the number 2 as "a metaphysical entity." But the class of couples, on the other hand, is indubitable and easily defined. It turns out, in fact, that the "class of all couples will *be* the number 2" (1956:542). So how self-evident is 1 + 1 = 2? When Whitehead and Russell (1927) set out to prove this proposition, it took them almost eight hundred pages to establish the basis for the actual demonstration. The proof is reached nearly one hundred pages into volume 2 of *Principia Mathematica* (1 + 1 = 2 is theorem #110.643). It took Leibniz, incidentally, only six short lines to prove 2 + 2 = 4. He considered 1 + 1 = 2 a statement of pure mathematics, true as a consequence of the law of contradiction and therefore true in all possible worlds.

For Plato, 1 + 1 = 2 describes relations that do not change between objects that do not change. It is independent of any preliminary constructive act; it reflects the reality of the Forms. Aristotle, by contrast, considered mathematics to be about idealizations that mathematicians construct (on the social contexts of these differences, see Restivo, 1983). For Kant, 1 + 1 = 2 is a synthetic proposition and a priori. Later the logicists, formalists, and intuitionists would offer mathematical arguments for different conceptions of 1 + 1 = 2. For the logicists, 1 + 1 = 2 can be expressed in terms of the logic of truth functions, the logic of quantification (in particular the concept of 'universal quantifier'), and the logic of classes (especially the concepts of 'sum class' and 'product class'). The statement "1 apple and 1 apple make 2 apples" is, just like 1 + 1 = 2, a statement of logic; it is not an empirical statement about *this* world, but rather a statement about "classes of classes in particular" (Korner, 1962:53).

For the formalists, 1 + 1 = 2 is an object and not a statement. Thus, as an object it is neither true nor false. But when it gets labeled in *their* reality as a "theorem-formula," the labeling can be viewed as a "true-or-false" phenomenon.

The intuitionists conceive 1 + 1 = 2 and one apple and one apple make two apples as "exact characteristics of self-evident, intuitive constructions" (Korner, 1962:178). It should be clear, then, that everyday arithmetic is not a simple matter for philosophers and mathematicians!

If we broaden our perspective somewhat, and at the same time leave the heights of *Principia Mathematica* and come down to a world of tuna fish, rocks, and cows, we can see why the apparently simple procedure of *adding* is *empirically* problematic. Consider the following problems from Davis and Hersh (1981:71–74; and see Hogben, 1940:32–34):

1. One can of tuna fish costs $1.05; how much do two cans of tuna fish cost?
2. A billion barrels of oil costs x dollars; how much will you have to pay for a trillion barrels of oil?
3. A banker computes your credit rating by allowing 2 points if you own your house, and then adds 1 point if you earn over $20,000 a year, and 1 point if you have not moved in the last five years; the banker subtracts 1 point if you have a criminal record, and 1 point if you are under 25, and so on. What does the final sum mean?
4. On an intelligence test, you get 1 point if you know George Washington was the first U.S. president; another point if you know some fact about polar bears, another point if you know

about Daylight Saving Time, and so on. What does the final sum mean?

5.   One cup of milk is added to one cup of popcorn. How many cups of the mixture will result?

6.   One person can paint a room in one day. He/she is joined by another person who can paint a room in two days. How long will it take for the two of them to paint a room?

7.   I have one rock that weighs one pound, and I find a second rock that weighs two pounds, How much will the two rocks weigh together?

These questions cannot be answered by simply following the imperative, Go ahead and add. You must know about the relationship between summing and discounting, dealing with a diminishing resource, "figures of merit," measurement problems, and so on. In problem 5, since a cup of popcorn can very nearly absorb a cup of milk without spilling, we could represent this as $1 + 1 = 1$. And in problem 7, we need to consider that weighing two rocks together can bring into play nonlinear spring displacements.

   Kline (1962:579–83) points out that we can object to the "truth" of $2 + 2 = 4$ on the grounds that the associative axiom is based on limited experience. But he notes further that problems such as those listed above show that there are even weaker links between arithmetic and "the real world." Consider, for example, the effect of supply and demand on the price of two herds of cattle sold separately and together; or the relationship between what is arithmetically correct (for example $2 \times 1/2 = 1$) versus what the case is in a particular instance (for example, do two half-sheets of paper make one whole sheet?); or adding forces that act at right angles to each other (in which case, for example, we could find that $4 + 3 = 5$). Kline's conclusion is not really the relativism his critic Gardner claims (see p. 3). His conclusion is that the system of $2 + 2 = 4$ arithmetic is based on limited and selected experiences. Ordinary arithmetic fails to describe correctly the results of what happens when gases combine by volume or one crop combines with another or when one cloud combines with another.

   There *are*, in fact, special arithmetics for dealing with special situations. Clocks that use the numerals 1 to 12 operate according to a modular arithmetic in which, for example, $10 + 6 = 4$. A finite group defined by its multiplication table according to a famous dictum by Cayley manifests the associative law but not the commutative law. Hamming (1980:89) uses an arithmetic and algebra in which $1 + 1 = 0$ (conventional integers are used as labels, and the real numbers are used as probabilities). In group theory, a set may be Abelian or non-Abelian according to whether the combining rule is commutative or non-commutative (cf. Wilder, 1981:39–40; Scriba, 1968:7).

We have not yet really reached the stage of an explicit social theory of mathematics, and yet there is plenty of reason already to ask questions about self-evidence. We have been alerted, however, to logic and self-evidence as cultural resources that can be used to defend or attack a social order (by Orwell and Dostoevsky, for example). Consider, furthermore, I. C. Jarvie's (1975) comment that nothing we would want to call mathematics or morality can be "localized;" there cannot be culture-bound answers on the question of whether children should be tortured or whether mathematical propositions are true or false. This explicit juxtaposition of mathematical and moral certainty is quite interesting. Let's see what happens when we try to put all of this into a sociological framework.

Mary Douglas writes that a self-evident statement is a statement "which carries its evidence within itself. It is true by virtue of the meaning of the words" (1975:277). Douglas takes her examples from Quine's discussion of self-evident sentences such as "all bachelors are unmarried men" and (no surprise here!) 2 + 2 = 4. Such sentences, Quine contends, "have a feel that everyone appreciates." People react to denials of such sentences the way they react to "ungrasped foreign sentences." Quine concludes that if analyticity intuitions operate substantially as he suggests they do, then "they will in general tend to set in where bewilderment sets in as to what the man who denies the sentence can be talking about" (Quine, 1960:66–67). Douglas improved Quine's account. Between the psychology of the individual and the public use of language, she inserted a dimension of social behavior in which logical relations also apply:

> Persons are included in or excluded from a given class, classes are ranked, parts are related to wholes . . . the intuition of the logic of these social experiences is the basis for finding the *a priori* in nature. The pattern of social relations is fraught with emotional power; great stakes are invested in their permanence by some, and then overthrown by others. This is the level of experience at which the gut reaction of bewilderment at an unintelligible sentence is strengthened by potential fury, shock and loathing. (Douglas, 1975:280)

The reason some of us can be so furious in identifying and opposing the illogical is that it is a threat to a *moral* order.

This prologue takes us to the threshold of a secular, earthbound view of mathematics as a social and cultural construct, product, and resource. This book is a contribution to an unfolding story of mathematics that is stripping it of the last clinging vestiges of Platonism and related forms of idealism. This book is not the end of the story, by any means. As readers will see, Platonism is not easy to uproot, even when the Platonist decides it is a good idea to embrace the idea of mathematics as a social practice. Nonetheless, this book is an important step forward. It provides some basic resources not only for grounding mathematical knowledge in mathematical practice but perhaps more importantly for linking it to the problem of improving the conditions under which we teach and learn mathematics and more generally the conditions under which we live.

INTRODUCTION

From at least the time of Plato, it has been customary to write stories about mathematics as if mathematics had fallen from the sky. This volume is a contribution to hauling "this lofty domain from the Olympian heights of pure mind to the common pastures where human beings toil and sweat" (Struik, 1986:280). The authors are all concerned with the bearing of mathematical practice on the production or construction of mathematics. But this volume does not outline a monolithic program. It is a portrait of struggles—with the ghost of Plato, for example, or with the spectre of mathematical practice. It is the editors' contention that these struggles are the starting point for the further development of a new understanding of mathematics already abroad, one that is grounded in social realities rather than metaphysical and psychological fictions. We therefore run the gamut from philosopher Michael Resnik's defense of a form of Platonism to my own sociological assault on philosophy and epistemology of mathematics. But we are not simply concerned with exploring a narrow band of philosophy and sociology of mathematics here. A great deal of space is devoted to issues of politics and values in mathematics and mathematics education.

Three general forms of mathematical studies are exhibited here: philosophy of mathematics, political and social theory of mathematics and mathematics education, and sociology and sociological history of mathematics. The opportunity for this undertaking was provided by the publication in 1988 and 1989 of special issues of two journals, *Philosophica* (edited by one of my coeditors for this volume, Jean Paul

Van Bendegem) and *Zentralblatt für Didaktik der Mathematik (ZDM)*, guest edited by my other coeditor, Roland Fischer). Van Bendegem brought together a group of students of mathematics, all of whom agreed about the need to pay attention to what real mathematicians can and actually do. We have selected four of the *Philosophica* contributions to publish in this volume.

Fischer called on a more diverse group of researchers and educators to assemble his two *ZDM* issues. These contributors were asked to address problems in the politics of mathematics education. Here, too, we find a concern for grounding our understanding of mathematics in mathematical practice. Selections from the *ZDM* issues appear in parts 3 and 4 of this volume. A number of these articles were originally published in German, and we are pleased to be able to publish them here for the first time in English translations.

Part 4 begins with a chapter by Fischer on mathematics and social change. This chapter reflects Fischer's editorial concerns in putting together the *ZDM* issues. The second chapter by Mehrtens is an English version of the paper he published in *ZDM* on the sociological history of mathematics under national socialism. The concluding chapter is my final word on the sociology of mathematics. This and the introduction to this volume are based on my contributions to the special issues of *ZDM* and *Philosophica*.

In the following pages, I briefly introduce the contributions to this volume and show how and to what extent they deal with the social realities of mathematical practice.

PHILOSOPHICAL PERSPECTIVES

The section on philosophical perspectives is introduced by coeditor Jean Paul Van Bendegem's paper. Just as Karl Marx demanded that social thinkers start their inquiries by looking at the real everyday activities of real people, so Van Bendegem argues that the study of mathematics must be grounded in mathematical practice—in particular, a theory or model of mathematical practice. Van Bendegem represents a small but growing group of philosophers of mathematics who are challenging traditional Platonist assumptions by asking questions such as, What are mathematicians really like, what *can* they really do, and what *do* they really do?

The fact is that pitting the spectre of mathematical practice against Platonism has not cleanly and quickly vanquished Platonism. To illustrate this point, we have included a paper by Michael Resnik, who advocates a form of Platonism. He finds himself in the position of having to confront the spectre of mathematical practice abroad in his field, but he meets the challenge not with a sociological tool kit and agenda but with a naturalistic strategy. Resnik's paper helps us to recognize the unresolved controversies alive in contemporary math studies. Even Van Bendegem, it should be noted, seems intent on preserving the possibility of something we could call *"the* mathematics." And while he wants to sociologize our understanding of mathematics, he envisages doing this in the form of theorems.

Resnik's goal is to show how a postulational account of mathematical knowledge could count as a naturalized epistemology. Here he is indebted to Quine's contention that epistemology is really a chapter in psychology and therefore a piece of natural science. Resnik appreciates the need to explain critically how we come to accept mathematics in terms of the everyday practices of mathematicians. His strategy is, however, not to sociologize mathematics but rather to naturalize Platonism. In the final analysis, his goal is to make pure mathematics a part of natural science.

Whereas Resnik addresses the increasingly obvious need for philosophers of mathematics to pay attention to mathematical practice, Thomas Tymoczko seeks to make mathematical practice the *focus* of the philosophy of mathematics. The major influence we see at work here is the quasi-empiricist imperative in the works of Imre Lakatos and Hilary Putnam. Quasi-empiricism involves an emphasis on mathematical practice and an orientation to mathematical methods as scientific methods. This approach links mathematics to scientific realism. In this view, mathematical objects, like the objects of scientific study, are considered "real." And mathematicians, like scientists, are considered to be discoverers—they make their discoveries in the realm of mathematical reality.

The emphasis on mathematical practice is designed in part to free philosophy of mathematics from all forms of foundational programs. But philosophers of mathematics who take mathematical practice into account are not unified on the questions of realism and Platonism.

Tymoczko's aim is to sort out some of the differences among philosophers of mathematics who come under the banner of quasi-empiricism. He focuses on contemporary set theory, treating it as one branch of mathematical practice rather than as a foundational program. He argues against adopting scientific realism in this case and suggests that perhaps some form of Putnam's "internal realism" is applicable. According to Tymoczko, there is no transfinite world "out there" waiting to be discovered. He stops short of the sort of analysis that would make sociological sense out of the concept of 'internal realism.'

Yehuda Rav describes the philosophy of mathematics as the study of the nature of mathematics, its methodological problems, its relation to reality, and its applicability. He notes the marked drift in the current literature towards an analysis of mathematical practice. He applauds this as a way to liberate philosophy of mathematics from Platonism, logicism, intuitionism, and formalism. Rav explores evolutionary epistemology as an approach that would ground mathematical studies in practice and simultaneously keep these studies abreast of the latest developments in the philosophy of science. Evolutionary epistemology, according to Rav, offers us a way to escape the "quicksand of neo-scholasticism and its offshoots."

Following Donald Campbell, Rav argues that, minimally, an evolutionary epistemology would be an epistemology that recognizes and is compatible with the biological and social evolution of humans. He argues furthermore that evolution—even biological evolution—is a knowledge process. The natural selection paradigm for knowledge evolution can be generalized to the cases of learning, thought, and science and also to mathematics. Rav argues, finally, that Platonism is completely incompatible with evolutionary epistemology.

This first set of papers illustrates that the spectre of mathematical practice is abroad in the philosophy of mathematics, but that it has not yet vanquished the ghosts of foundationalism and Platonism. And, so far, the efforts in this direction seem generally to lead to forms of naturalism rather than to sociological theories. In part 3, we explore approaches to mathematics studies that still fall short of radically sociologizing mathematics but that are more firmly grounded in the politics of mathematics and mathematics education.

MATHEMATICS, POLITICS, AND PEDAGOGY

Part 3 is introduced by coeditor Roland Fischer's paper on mathematics as a means and as a system. Set theory plays a prominent role in his argument. Set theory, Fischer claims, does not allow us to define elements in a set in terms of their relationships with all the other elements in a set. This notion, he argues, is mirrored in "rational man" economic theory. He then points out the crucial flaw in "rational man" theories: the failure to recognize a fundamental sociological thesis propounded by Karl Marx that the individual is really an ensemble of social relationships.

Fischer defines a field called "didactics of mathematics," a collective effort to study and shape the relationship between human beings (individuals, groups, and whole societies) and mathematics. Fischer views mathematics as a *means* for people, a resource, *and* a *system* of concepts, algorithms, and rules embodied in ourselves, our thinking, and our actions. Fischer pursues these themes further in the first chapter of part 4. Here he makes quite clear his humanistic concerns, and the importance of collective self-reflection for improving the human condition. He argues that mathematics is a mirror of humanity and can contribute to the process of collective self-reflection. He stresses the importance of valuing freedom, autonomy, flexibility, and playfulness in social affairs and mathematics. Given its potential as a factor in and for humanistic social changes, mathematics should not be permitted to be totally focused on its traditional tasks.

Helga Jungwirth's paper illustrates the inability of prevailing approaches and models in math studies to reconstruct "mathematical cultures" in ways that reveal the contexts of our relations to mathematics. Her objective is to identify the underlying assumptions and limitations of research on women and mathematics. She argues that these assumptions and limitations reflect the lack of systemic-ecological contextualist models in math studies. The result is a correspondence between claims about discovering the specific relations between women and math, the goal of changing those relations, a psychological approach to the problem, and the use of empirical-analytical methods. Jungwirth claims that the social realities of mathematical practice are overlooked, and that stereotypical thinking about gender as well as an uncritical view of mathematics dominate the conventional paradigm. In order to see what sorts of changes we need to look toward to solve the problems Jungwirth identifies, we should look to papers like the next one by Nel Noddings.

Noddings argues forcefully that mathematics classrooms should be politicized. She integrates Paolo Friere's pedagogy and a feminist perspective to ground her contention that students should be involved in planning, challenging, negotiating, and evaluating the work they do in learning mathematics. Such involvement, in fact, can be expected to facilitate math learning.

One of the core approaches to contemporary math pedagogy is constructivism. Constructivists in math education argue that all mental acts (perceptual and cognitive) are constructed (this is not the same sort of constructivism found in the social problems literature or in social studies of scientific knowledge). Noddings's argument is that constructivism, assuming it is a viable pedagogy (an assumption coming under increasing scrutiny and criticism), must be embedded in an ethical or political framework if it is going to have an effect on reforming classroom practice.

Noddings links what goes on in the mathematics classroom to education for civic life in a free society. She is therefore necessarily critical not only of math pedagogy but of modes of schooling in general. The primary aim of every teacher must be not simply to promote some uncritical notion of achievement but the growth of students as self-affirming and responsible people.

The theme of political dimensions in mathematics education is pursued further by Ole Skovsmose. He criticizes the focus of mathematics pedagogy on the individual learner and is thus necessarily critical of Piaget's genetic epistemology (widely influential among science educators). Like Noddings, Skovsmose wants us to explore the epistemological potential of the relationships between the children in a classroom. Skovsmose's paper illustrates that one of the paths to a focus on social interaction and social practice starts from Wittgenstein's work in the philosophy of language and Austin's theory of speech acts.

In the end Skovsmose asks us to rethink the concept of 'knowledge' so that Platonic dreams are replaced by realistic ideas about knowledge conflicts and reflexivity that make it possible to evaluate technologies.

The final contributor to part 3 is Philip Davis, whose work combines the premise that mathematics is social practice, the imperative that mathematical practice be made the object of our descriptive and interpretive strategies, and the assumption that we live—and must learn to live—in a mathematized world. Mathematical education must, Davis argues, be reoriented to help us find the right ways to understand mathematical practice.

MATHEMATICS, SOCIETY AND SOCIAL CHANGE

In the lead article for part 4, coeditor Roland Fischer reinforces some of the ideas introduced in part 3 by arguing for an orientation to the tasks that face humanity today, the importance of a strategy of collective self-reflection for undertaking those tasks, and the role of mathematics in fostering self-reflection. Fischer draws on the ideas of the sociologist Niklas Luhmann, who has dealt with the problem of complexity from an epistemological standpoint. Fischer is concerned to show how mathematics can help us deal with establishing and understanding the complex interconnections that characterize our world. He also stresses that mathematics itself is an expression of social relations, an idea developed more fully in my concluding essay in this volume.

The second paper in part 4, by Herbert Mehrtens, in an example of a sociologically grounded history of mathematics. Luhmann's work once again enters the picture by providing Mehrtens with a conception of 'social system.' Mehrtens's paper is a survey of problems arising in German mathematics under national socialism. He begins by describing the social system of mathematics in Germany during the rise and rule of national socialism, and the characteristics of national socialism. He then analyzes the background and structure of the attempt to construct a "German" mathematics related to Nazi ideology. Mehrtens goes on to examine the basic relations between mathematical and political thought. The focus here is on the twin processes of adaptation and resistance by professional societies. Mehrtens argues that social differentiation within the system of mathematics, as well as its modern cognitive and social universality, were preconditions of adaptations.

In the concluding chapter in part 4, I outline a radically sociological approach to thinking about mathematics. The approach is radical because it claims complete jurisdiction over the problems of the nature of mathematics and mathematical knowledge for sociology. It is not my intention to simply be political. The rationale for such an approach has been developing since the 1850s. Once we realized that individual human beings are in fact *social* beings, it was only a matter of time before it occurred to some social theorists that the mind and thinking are social phenomena. The upshot of this sociological revolution is the argument and sketch in the concluding essay.

CONCLUSION

The idea that mathematics, or any other form of knowledge, falls from the sky is quickly fading. But sometimes, as in the case of our ideas about the gods, the difficulty of coming up with a satisfactory alternative explanation keeps the old idea alive. That is the case in mathematical studies today. Some students of mathematics recognize that there has been progress in social studies of mathematics, and they take up the rhetoric of culture, communication, and community in their analyses of mathematics. Others adapt a fashionable (for good reasons, let me add) interdisciplinary approach, but play this game close to the chest so that traditional disciplinary boundaries and assumptions are not seriously threatened. Others rush into social constructivism with such abandon that they are almost at a loss for words when it comes to exchanges with less adventurous colleagues. But within all of this diversity, there is a growing awareness, if not of a theoretical social constructivism, at least of the necessity of attending to the social practices that people engage in to produce or construct knowledge and facts, including mathematical knowledge and facts.

We are in the early stages of transforming insights from the works of Durkheim, Neitzsche, and others into collective representations. This volume is one small step in support of that transformation.

REFERENCES

Davis, P. J., and R. Hersh
  1981 *The Mathematical Experience.* Boston: Houghton Mifflin.

Dostoevski, F.
  N.d. "Notes From Underground," pp. 107–240 in *The Best Short Stories of Dostoevsky.* New York: Modern Library.

Douglas, M.
  1975 *Implicit Meanings.* London: Routledge and Kegan Paul.

Gardner, M.
  1981 "Is Mathematics for Real?" Review of P.J. Davis and R. Hersh, *The Mathematical Experience. The New York Review of Books* 28:37–40.

Hamming, R. W.
  1980 "The Unreasonable Effectiveness of Mathematics." *American Mathematical Monthly* 87 (February) :81–90.

Hardy, T.
  1969 *Jude the Obscure.* New York: Bantam.

Hogben, L.
  1940 *Mathematics for the Million.* Rev. and enl.. New York: W. W. Norton.

Jarvie, I. C.
  1975 "Cultural Relativism Again." *Philosophy of the Social Sciences* 5:343–53.

Jourdain, P.
  1956 "The Nature of Mathematics." Pp. 4–72 in J. R. Newman (ed.), *The World of Mathematics*, vol 1. New York: Simon and Schuster.

Kline, M.
  1962 *Mathematics: A Cultural Approach.* Reading, Mass.: Addison-Wesley.

1980 *Mathematics: The Loss of Certainty*. New York: Oxford University Press.

Körner, S.
1962 *The Philosophy of Mathematics*. New York: Harper Torchbooks.

Mannheim, K.
1936 *Ideology and Utopia*. London: Routledge and Kegan Paul.

Orwell, G.
1956 *1984*. Baltimore: Penguin Books.

Quine, W. V. O.
1960 *Word and Object*. Cambridge, Mass.: MIT Press.

Restivo, S.
1983 *The Social Relations of Physics, Mysticism, and Mathematics*. Dordrecht: D. Reidel.

Russell, B.
1956 "Definition of Number." Pp. 537–43 in J. R. Newman (ed.), *The World of Mathematics*, vol 1. New York: Simon and Schuster.

Scriba, C. J.
1968 *The Concept of Number*. Mannheim: Bibliographisches Institute.

Spengler, O.
1926 *The Decline of the West*. Vol. 1. New York: International Publishers.

Struik, D.
1949 "Mathematics." Pp. 125–52 in W. Sellars, V. McGill, and M. Barber (eds.), *Philosophy for the Future*. New York: Macmillan.
1986 "The Sociology of Mathematics Revisited: a Personal Note." *Science & Society* 50 (Fall) :280–99.

Whitehead, A., and B. Russell
1927 *Principia Mathematica*. Cambridge: Cambridge University Press.

Wilder, R. L.
1981 *Mathematics as a Cultural System*. New York: Pergamon.

PART II: PHILOSOPHICAL PERSPECTIVES

2

Foundations of Mathematics or Mathematical Practice: Is One Forced to Choose?

INTRODUCTION

Philosophers of mathematics can be roughly divided into two groups. Type I is particularly fond of questions such as: What are *the* foundations of mathematics? What are numbers? What is a set? What is Church's thesis really about? What is decidability? What is infinity? What is mathematical truth? These questions are all situated within mathematics proper. Formalists, logicists, intuitionists, constructivists, and finitists (strict and otherwise), are in this sense Type I. Type II, however, wants answers to questions such as: How is mathematics done? What is a *real* mathematical proof? Why do mathematicians make such a fuss over the use of computers in order to find and construct proofs? Is it possible to gather evidence concerning the plausibility of the correctness of a mathematical statement? How is it possible that an accepted proof turns out to be wrong? Type II is still a rare species but happily enough—that is, if you happen to be Type II as well—this is changing.[1] But it would be an exaggeration to claim that something like a theory of mathematical practice, mathematics as it is done, exists. There are plenty of ideas, plenty of detailed studies, but there is no general framework. I take it that hardly any argument, in fact none, is needed to show the importance of such a theory. If you wish to study problems related to the educational aspects of mathematics, or the diverse and complex relations between mathematics and the culture at large, or the psychological and social processes of mathematical invention and construction, you will obviously need a theory or at least a model of what mathematical practice is about.

In this short chapter I do not intend or pretend to present (the outline of) such a theory or model. My aim is quite modest, although the point I wish to present is, philosophically speaking, an important one. In the search for this model or theory of mathematical practice, most Type II researchers seem to agree that models and theories used by Type I philosophers of mathematics are not interesting. After all, their approach is a highly normative one, ignoring all aspects of real mathematical life. Either these models are criticized or they are just simply ignored. My point is that, although acknowledging that Type I and Type II researchers are actually in different fields, their theories and models are, to a large extent, commensurable (to use a fashionable term). The basis allowing for the possibility of commensurability is constituted by the notion of an *artificial mathematician*. In Type I research, there are plenty of artificial mathematicians around. The two most famous ones are Hilbert's ideal mathematician and Brouwer's creative subject. In Type II research, we obviously are talking about real mathematicians. It is therefore a natural question to ask whether real and artificial mathematicians are related. And, if so, can these possible relations form the background against which to compare Type I and Type II theories. As will be shown, there is a gradual transition from extreme Type I theories to extreme Type II theories.

THE GODLIKE MATHEMATICIAN

No doubt most working mathematicians assume set theory— that is, ZFC, Zermelo-Fraenkel set theory with the Axiom of Choice— is the best (Type I) foundation around for mathematics at the present moment. The standard formulation consists of (1) some version of classical first-order logic and (2) the typical set-theoretical axioms. In such a foundational theory no mention is made of a mathematician. The set-theorist will (rightly) claim that the logical axioms and rules mention only the logical signs and the set axioms mention only sets and operations on sets. However, the fact that no properties of a mathematician are listed implicitly or explicitly in the theory does not imply that therefore the theory deals only with mathematics and not with mathematicians. A straightforward way to associate a mathematician with a mathematical theory is quite simply to ask the following question: supposing there is a being that has the property that it knows everything that the mathematical theory claims, what property does that being have? Note that the question is a trivial one if asked in a Type II approach. In that case, one starts with the mathematician (or the mathematical community) and then studies how the mathematician does mathematics. The question is less trivial when asked in a Type I context. In order to clarify this strategy, let me present a first example.

Most mathematicians would agree with the following statements: (1) there is something like a mathematical universe, (2) this universe is unique, and (3) in it all mathematical problems are settled. The mathematician's task is to discover and chart this universe, with the knowledge that a complete map is impossible. But suppose that there is a being with the property that it has full knowledge of the mathematical universe. What epistemic properties does this being possess? Two important properties immediately follow. First, its knowledge is strongly complete. By (3), all mathematical problems are settled; therefore, given a mathematical problem or statement A, either A is the case in the mathematical universe or not-A is the case in the mathematical universe. Secondly, by (1) and (2), its knowledge is weakly complete as well. In model-theoretic terms, (1) guarantees the existence of a model, whereas (2) guarantees the uniqueness of this model. If this hypothetical mathematician has full knowledge of this model, this obviously implies the weak completeness. From these two properties, a third, crucial one is derived: this being must have truly godlike powers! The reason is quite simple. For an epistemic subject, to know a strongly and weakly complete first-order theory implies it must have an actual infinite capacity to store knowledge. If the capacity is restricted to potential infinity, then undecidability results become unavoidable and full knowledge of the mathematical universe is no longer possible. Errett Bishop summarized his critique of classical mathematics when he wrote: "classical mathematics concerns itself with operations that can be carried out by God" and "if God has mathematics of his own that needs to be done, let him do it himself" (1967:2). In terms of the above analysis, an even stronger statement can be made: he is the only one who can do it; he has to do it himself.

It is perhaps interesting to present an example of the epistemic strength of this God-mathematician (GM). Bishop (1985)[2] himself introduced the following example. Let (An) be a binary sequence. Then the GM will accept the following principle, the so-called *Limited Principle of Omniscience* (LPO): Either there is an n such that $An = 1$, or else $An = 0$ for all n. If we assume that GM can indeed decide this problem, then he can solve the following problem. Take an unsolved mathematical problem, for example, Fermat's Last Theorem (FLT) or Goldbach's Conjecture. Now consider the following sequence : in the sequence (An), $An = 1$ if FLT is provable and $An = 0$ if not-FLT is provable. Obviously the sequence (An) will consist either of all ones or of all zeros. However it is not obvious at all which one is the case (at least for mortals). Yet, if LPO holds, GM can decide the matter. Thus GM can decide whether FLT or not-FLT holds. Note that for GM, mathematics ceases to be an interesting enterprise, for the simple reason that everything is already known.

It is interesting to note the close similarity between GM and the Demon of Laplace. In the very same sense that Laplace's Demon corresponds to the ideal physicist, GM corresponds to the ideal mathematician. For the Demon, too, the universe ceases to be an interesting place, as it holds no secrets. For the Demon, too, time ceases to be real, just as GM lives in a timeless realm. One might well wonder whether the parallel breakdowns of the Demon and GM are related or not.

THE CONSTRUCTIVIST MATHEMATICIAN

If godlike mathematicians have little or nothing to do with us, are we not best advised to scale this hypothetical being down to our size? Basically, there are two options: (1) assume the existence of a unique, mathematical universe, but deny one can have a full knowledge of it, and (2) deny the existence of a unique mathematical universe altogether. The second option corresponds roughly to the route taken by Brouwer, whereas the first option is currently explored in epistemic mathematics.[3] The crucial difference between these two approaches is directly linked to the discovery-construction distinction. Are we wandering around in a mathematical universe wherein we discover mathematical theorems, or are we just exploring a creation of our own making? I will not go into this discussion—this is a quite separate topic—for it is sufficient to note that in both cases the answer to the following question will be the same: what is the epistemic content of a hypothetical mathematician whose capacity is limited to potential infinity? The answer is: what is accessible to the mathematician on the basis of construction and proof. Although perhaps at first sight this answer may seem a clear one, it is nevertheless highly ambiguous. The history of (the philosophy of) mathematics has show us that there are many different ways to sharpen this answer. In other words, there are many constructivist mathematics imaginable. However, as I will argue, they all share a set of nonhuman properties. Or, to put it differently, a constructivist mathematician (CM) still has some distinctly Type I properties that distinguish it clearly from a Type II mathematician. Thus the differences do not appear to be essential for my argument. Nevertheless, let me briefly present three examples to illustrate the richness of the constructivist approach.

For the intuitionist, CM(I) knows A if there is, in principle, a proof or construction of A available, that is, CM(I) is capable of producing a proof of A, or a construction for A. Obviously, CM(I) will reject LPO. But CM(I) will also reject Markov's principle (MP): If (An) is a binary sequence such that it is not the case, for all n, that $An = 0$, then there is an n such that $An = 1$. The reason is that for the intuitionist not-A means that given a proof or construction of A, this proof or construction can be extended into a proof of something absurd or into an impossible construction. Thus not-A stands for "if A, then absurdity." In the case of MP, if CM(I) has shown that it is not the case that for all n, $An = 0$, then he or she has only shown that the assumption that all $An = 0$ leads to an absurdity. This gives him/her no clue as to how the n, such that $An = 1$, can be found or constructed.

The Russian constructivist, CM(R), however, accepts MP. The reason here is that the notion of construction is replaced by the notion of algorithm in an extended sense. Cases such that, on the one hand, one knows that the algorithm will end on a certain input, but, on the other hand, no finite bound can be specified beforehand, are accepted. However, CM(R) will reject some intuitionist principles, such as the Fan Theorem (FT).

A third version is Bishop's constructivist, CM(B). This is the weakest version, as neither MP nor, for example, FT is accepted. The main advantage of Bishop's constructivism is that it is consistent with classical analysis (assuming the consistency of the latter, of course). Both intuitionist and Russian constructivism are extensions of Bishop's constructivism, but both are inconsistent with classical analysis. Furthermore, intuitionism is inconsistent with Russian constructivism. Note, too, that these three approaches do not exhaust the whole range of constructivist theories. I refer the reader to Beeson (1985) for an overview. Let me now return to the main line of the argument. What properties of CM(x)—where x is your favorite brand of constructivism—are still clearly of Type I? Basically, there are two aspects of prime importance. Actually, these two problems will appear only too familiar to anyone acquainted with epistemic logic.[4]

The first problem has to do with true knowledge. If CM(x) knows A, then A must be the case. In other words, the case wherein CM(x) knows A, but not-A is a mathematical theorem, does not occur.

I will refer to this principle as the *principle of Immunity of Error* IE. It is hardly necessary to argue that IE is a typical Type I property. Real mathematicians do believe impossible things from time to time. They did, for example, believe such nonsensical statements as $(\sqrt{-1})^2 = -1$. Moreover they knew that these statements were nonsensical. The fact that these mathematicians were fully aware of the absurdity involved shows that an argument of the following type does not apply. One might propose to weaken the IE. Instead one could adopt the principle wIE (weak Immunity *of Error*): If one knows that one knows A, then A must be the case. In other words, just knowing A does not guarantee the correctness of A. But that does not work. And it is useless to weaken wIE even further, for what meaning could be given to the statement that One knows that one knows that one knows that A without it being the fact that one knows that one knows that A? Furthermore, they managed to deal with these absurdities and to derive interesting, important, and, above all, correct mathematical conclusions from them. To quote another famous historical example, Berkeley did show convincingly that Newton's treatment of infinitesimals was inconsistent, but most historians will agree that it was a good thing for the development of mathematics, analysis in particular, that Newton largely ignored this criticism and continued to develop this inconsistent theory. Actually, with the advent of nonstandard analysis, one could argue that consistent talk about infinitesimals is, in fact, possible.

The second problem has to do with the *principle of Immediate Consequences* IC. Suppose that CM(x) knows A and that B is a logical consequence of A. Then CM(x) must also know B. IC is surely acceptable, for it says nothing but: if you know A and there is a proof—according to your favorite brand x—of B from A, then surely you must know B. But if this is acceptable, then it has the immediate, startling conclusion that if CM(x) knows A, then CM(x) must know *all* logical consequences from A. And this seems less or not at all acceptable when discussing real mathematicians. Obviously no real mathematician has such insight. I mentioned earlier ZFC as the foundation used today by most working mathematicians. Every mathematician who knows these axioms, therefore knows all the logical consequences of these axioms, that is, he/she knows all the theorems of set theory. One might object that for CM(x) to know that B is a logical consequence of A means that CM(x) has a proof *in principle* of B from A. Thus, to know a logical consequence means quite simply to be able to present a proof when asked to do so. But this only increases the mystery: what kind of knowledge is this knowledge of "proofs in principle"? One way or another, this must reduce to having direct access to the mathematical universe, where one can "see" whether B is a logical consequence of A or not. True, CM(x) can see only part of the universe; nevertheless, it is somewhat startling to come to the conclusion that GM and CM(x) are closer relatives than one might have imagined.

THE FINITIST MATHEMATICIAN

How should we modify CM(x) such that the IE principle and the IC principle no longer hold? In order to reject the IC principle, it is sufficient to replace the notion of proof *in principle* by the notion of *real* proof. A real proof is characterized by the fact that it should be recognizable by a mathematician as a proof that is bounded in time and in space. Real proofs are sequences of signs written in some language or other. It seems appropriate to call a mathematician thus limited a finite mathematician (FM). Actually, in this case, too, it would be better to speak of FM(x) for, as Ernst Welti (1987a, 1987b) has shown in his excellent historical study, there are many types of finitist mathematics, strict or otherwise, around. However, just as in the constructivist's case, it is not necessary to go into details. It is easy enough to see that the presence of finite bounds must result in the violation of the IC principle. For suppose, to keep matters simple, that an overall bound, say L, is defined on the length of proofs. FM can only check and thus accept or reject proofs below a certain upper bound. Suppose further that FM has accepted A as a theorem after inspecting the proof (having a length less than L) of A. Finally, suppose that FM has also accepted a proof of "if A, then B," and that this proof also has a length less than L. It does not follow that therefore FM has to accept the proof of B, since the proof of B may have a length larger than L. For the proof of B will be the result of the concatenation of the proof of A and of the proof of "if A, then B." Thus it is rather easy to reject the IC principle. The IE principle, however, is a quite different problem.

On the one hand, it is obvious that the possibility of error should be allowed. The history of mathematics presents an interesting story of what one could call *creative* mistakes. Precisely because mistakes were made, the mathematical community was able to see the next step to take. But, on the other hand, it is not clear at all how one should proceed. What principles can be formulated about an artificial mathematician that allow this being to make mistakes *and* to learn from them? Two alternatives present themselves. The first one is to replace the underlying contradiction-free logic of mathematics by a paraconsistent or a dialectical logic (see Priest, Routley and Norman, 1989; Arruda, 1980). The possibility is then allowed for accepting that "if FM knows A, then A is the case," that "FM knows A," yet that "not-A is the case." However, this first alternative will surely have to be supplemented by some methods for "repairing" the error. But, as must be obvious, these methods cannot be algorithms. If they were, it would be sufficient to apply them each time a contradiction arises, thus establishing a modified form of the IE principle. If an error occurs, it can be "calculated away." Thus heuristics have to be introduced. Innocent though this conclusion may seem, it is of fundamental importance. So far, we always assumed that whatever the artificial mathematician learns about the mathematical universe, it is learned truthfully. At this point, the possibility is introduced that the artificial mathematician may be misled by what he or she thinks to be the case in the mathematical universe. In other words, this universe itself can no longer be used as a justificative device. FM can no longer say, "I believe or I know A, because A is a mathematical fact and, therefore, true in the mathematical universe." FM will have to look for other criteria to convince himself or herself, that he or she knows A truthfully.

THE REAL INDIVIDUAL MATHEMATICIAN

The Type I philosopher might remark at this point that it is clearly impossible to have mathematicians making errors. If we restrict ourselves to FM-like mathematicians, then any proof presented will be a surveyable proof because of the limits imposed on time and place resources. But a surveyable proof can decidably be found out to be error free or not. If not, the error can be located and repaired. Why, then, do we need the heuristics? The answer, in all its simplicity, is this: what most mathematicians write and read most of the time are *not* proofs, in the formal sense of the word. They are what I have called elsewhere "proof-outlines" (Van Bendegem, 1988). That is, what are presented are the major steps in the proof. The mathematician who writes the proof thus assumes that a trained mathematician, with sufficient knowledge of the particular mathematical field the proof is about, is capable of filling in the missing steps. The problem is not that mathematicians should be accused of laziness or sloppiness, the matter is quite simply that the demand of, formally speaking, correct proofs, is an impossible one. If something has been made clear by *Principia Mathematica*, then surely it is the fact that that is not the way to do mathematics. If errors occur in a proof-outline, they do so because the mathematician assumed wrongly that a particular step could be filled in. Therefore, errors are likely to occur—the history of mathematics tells us so[5]—and heuristics are needed to repair these errors.

To illustrate this thesis, let me present three heuristics that have been frequently employed in mathematics to search for errors and to repair the damage if errors occurred.

The first example is well known from Lakatos's brilliant study (in *Proofs and Refutations*) of Euler's conjecture, $V - E + F = 2$, that is, the statement that, given a polyhedron, the number of vertices minus the number of edges plus the number of faces always equals two. The three-part heuristic Lakatos arrives at is the following:

RULE 1. *If you have a conjecture, set out to prove it and to refute it. Inspect the proof carefully to prepare a list of non-trivial lemmas (proof-analysis); find counterexamples both to the conjecture (global counterexamples) and to the suspect lemmas (local counterexamples).*

RULE 2. *If you have a global counterexample discard your conjecture, add to your proof-analysis a suitable lemma that will be refuted by the counterexample, and replace the discarded conjecture by an improved one that incorporates that lemma as a condition. Do not allow a refutation to be dismissed as a monster. Try to make all "hidden lemmas" explicit.*

RULE 3. *If you have a local counterexample, check to see whether it is not also a global counterexample. If it is, you can easily apply Rule 2. (Lakatos, 1976:50)*

The second example concerns a heuristic that I have labelled "confining inconsistencies" (Van Bendegem, 1985). In the pre-Newtonian and pre-Leibnizian period in the development of analysis, many mathematicians—Giles Personne de Roberval, John Wallis, and François Viète, to name but a few—were developing mathematical theories that were clearly inconsistent. However, they did manage to work with these inconsistencies because in many cases these inconsistencies were confined. An example may clarify the matter. Wallis had the following beautiful proof for the area of a triangle (figure 2–1).[6] The triangle is divided into an infinite number, ∞, of lines. The area of the triangle is the sum of all the tiny rectangles. Each of these rectangles has a height H/∞ and a length b. Thus the area of the triangle is equal to $\sum b.H/\infty$ or $H/\infty.\sum b$. $\sum b$ is an arithmetical progression with an infinite number of terms, with first term B and last term 0. Hence $\sum b = B.\infty/2$. Thus the area is equal to $H/\infty.B.\infty/2 = H.B/2$. Wallis knew

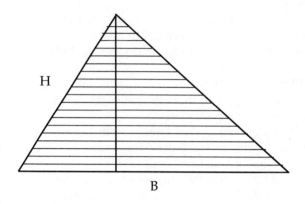

Figure 2–1.
Wallis' Proof for the Area of a Triangle.

that was the result he should obtain, because there are many ways (and different methods) to obtain the area of a triangle. Although the proof seems hilarious—at least to modern eyes—the conclusion is correct. Therefore in any other proof in which the area of a triangle had to be caluclated, Wallis could safely insert the above proof. In this sense, the inconsistency is confined; it is not allowed, so to speak, to escape from the proof wherein it occurs.[7]

The third example is a heuristic so familiar to mathematics that probably most mathematicians are not aware of the crucial role it plays: *multiple proofs* or proof-outlines. Mathematicians do spend a lot of their time rewriting proofs and searching for different proofs of theorems that already have been proved. For some famous theorems, the list is quite impressive: there are no fewer than ninety-six different proofs of the Pythagorian theorem (Versluys, 1914). Every mathematician knows at least two proofs of the existence of an infinite number of primes (the classical Euclidean proof and the proof related to the Riemann zeta-function). I have shown elsewhere how this heuristic plays an important role in evaluating the importance of a mathematical problem (Van Bendegem, 1987).

I must emphasize that the heuristics presented here are clearly *rough* heuristics. True, they are more specific than, for example, Polya's first heuristics in *How To Solve It*. But tasks such as "Try to prove the conjecture" and "Find an alternative proof" are not exactly helpful for the working mathematician. What is needed is a worked-out theory of more specific, domain-related detailed heuristics. In the area of Automated Reasoning, one of the many branches of Artificial

Intelligence, this is precisely what one is looking for. However, the way things look at the present moment, it is still the case that only formal proofs are considered. To my knowledge, no work is being done on the level of proof-outlines. On the other hand, it must be mentioned that within the computer world, nonmonotonic logic is highly developed. This logic enables its user to revise currently held beliefs. In this sense, it shares some properties with paraconsistent and dialectical logics, although it is very different from them (see Ginsberg, 1987). As I indicated in my discussion of the finitist mathematician, this is certainly needed as a crucial element in any theory of the real individual mathematician.

THE REAL SOCIAL MATHEMATICIAN

Individual heuristics do not, however, tell the whole story. Of course, it would be an easy way out to claim that the mathematician, as part of the mathematical community, exists *as a mathematician* only by virtue of his or her membership in that community. But that does not explain why, if one is interested in understanding the dynamics of mathematical change (as Type II philosophers are), social elements should be taken into account. As in the case of the use of heuristics, I believe there to be at least two major arguments in support of this thesis.

The first argument relates to a point argued for in the preceding paragraph. Mathematicians do not write proofs but proof-outlines. Proof-outlines do not have a standard form in the sense that formal proofs do. They are not a sequence of formulas, in which each formula is either an axiom or the result of the application of a derivation rule on formulas already occurring in the list. Instead they can take many different forms. As an example, compare these two proofs of the same theorem, namely the fundamental theorem of arithmetic.[8]

Version 1. To prove the result, note first that if a prime p divides a product mn of natural numbers then either p divides m or p divides n. Indeed if p does not divide m then $(p, m) = 1$ whence there exist integers x, y such that $px + my = 1$; thus we have $pnx + mny = n$ and hence p divides n. More generally we conclude that if p divides $n_1 n_2 \ldots n_k$, then p divides n_1 for some ι. Now suppose that, apart from the factorization $N = p_1^{j_i} \ldots p_k^{j_k}$ derived above, there is another decomposition and that p is one of the primes occurring therein. From the preceding conclusion we obtain $p = p_1$ for some ι. Hence we deduce that, if the standard factorization for N/p is unique, then so also is that for N. The fundamental theorem follows by induction.

Version 2. First, N must have at least one representation, $N = p_1^{a_1} p_2^{a_2} \ldots p_n^{a_n}$. (1) Let a be the *smallest* divisor of N that is > 1. It must be prime, since if not, a would have a divisor > 1 and $< a$. This divisor, $< a$, would divide N and this contradicts the definition of a. Write a now as p_1, and the quotient N/p_1 as N_1. Repeat the process with N_1. The process must terminate, since $N > N_1 > N_2 > \ldots > 1$. This generates eq. (1). Now if there were a second representation, by the corollary of theorem 6, each p_i must equal some q_i, since p_i/N. Likewise each q_i must equal some p_i. Therefore $p_i = q_i$ and $m = n$. If $b_i > a_i$, divide $p_i a_i$ into two equations, C. The second representation is $q_1^{b_1} q_2^{b_2} \ldots q_n^{b_n}$ (2). Then p_i would divide the quotient in eq. (2) but not in eq. (1). This contradiction shows that $a_i = b_i$.

These two proofs are sufficiently different to warrant the introduction of the notion of *style* in mathematics. It is not an exaggeration to claim that a mathematician develops a certain type of style and that one can identify him or her by it. It also implies—and here the social element enters the picture—that mathematicians sharing the same style will understand each other better. After all, they do speak the same language, or, one should say, the same mathematical dialect. Seen from this perspective, the Bourbaki project, apart from its mathematical content, was a project equally important in its proposal for a new mathematical style. The Bourbaki volumes aspired to be a new foundations of mathematics, but, at the same time, they constituted a manual of style for it.

The second argument has to do with a recent, intertwined, twofold development or, better, change in mathematical practice.

Long proofs are not uncommon in mathematics, as is well known. However, it is a quite recent phenomenon that some proofs turn out to be so long that an individual mathematician is incapable of surveying them. The exemplar in this case is the classification theorem of finite groups, estimated at about fifteen thousand pages (Gorenstein, 1986). Instead, one can claim only that the proof is socially surveyable, not individually surveyable. Mathematician A has checked part X and mathematician B part Y, and, putting their efforts together, they come to the conclusion that the whole proof is correct. Neither A nor B individually can make this claim, but together they can. Or, in other words, the proof *as a mathematically accepted proof*, exists only on the social level. Hence, the basic unit to consider is not the individual mathematician but the mathematical community.

The related part has to do with computer proofs. Since the "drama" of the four-color theorem, it has become apparent that the presence of the computer as a symbol-manipulating device, must have its effect on mathematical practice. If part of the proof has been carried out by computer, and if the calculations are so cumbersome and intricate that neither a human mathematician nor the mathematical community is likely to check it in detail, are we then in a position to accept the proof? If one is tempted to answer yes to this question, then one must accept the conclusion that "proof," as classically understood, is not the only way to establish new mathematical results. This is really going beyond heuristics, for heuristics points the way to a classical proof, whereas, in the computer case, the computer calculations are the best available. The problem is not a recent one. Mark Steiner (1975) has already made a case for other methods, besides mathematical proof (in the classical sense, that is, up to the real individual mathematician [RIM]), to establish the truth of a mathematical proposition. Perhaps one is not inclined to follow along such a route, but, if understanding mathematical practice is the goal, these aspects will have to be taken into account.

Although the real social mathematician (RSM) is not the end of the continuum—surely we should go further and consider the real social mathematician in society at large—I hope I have made a convincing case for the idea that GM and RSM, although worlds apart, are related.

A TENTATIVE CONCLUSION

The subject of this chapter, basically, was to answer this question, If X is any type of mathematician, then for X to know A, where A is a mathematical statement, means exactly what? We have progressed from the godlike mathematician, GM, for whom the answer was quite straightforward. For GM to know A, is simply equivalent to A being true in the unique mathematical universe. Along come the constructivists who want to scale down GM to some kind of "ideally real" mathematician, CM(x). CM(x) knows A if CM(x) has a proof or construction available of A, in principle. Replacing the "in principle" part by "actually," CM(x) is transformed into some kind of finitist mathematician, FM. But—relying on some well-known arguments about epistemic logic—as it turns out, even FM is still a highly idealized being. Establishing a link with the real—individual or social—mathematician forces us to introduce elements into the story that one would perhaps not expect in the mathematical context: heuristics, failure, and error (and therefore revision), style (and therefore aesthetics) and the social coherence of the mathematical community.

Although it is clear that the transition from GM to RSM is a gradual one, the differences between the extremes of the continuum are enormous. But that has been known all along. The more interesting part is that it is gradual. Seen from the viewpoint of the RIM or the RSM, then the FM, CM(x), and GM are to be seen as increasing and, therefore, simplifying and helpful abstractions. Note too that all mathematicians mentioned, artificial and otherwise, are only snapshots from an immense gallery of possibilities.

Perhaps the reader wonders why I am so insistent on this point. Basically, there are two reasons. The first reason has to do with our understanding of RIM and RSM. As said, FM, CM(x), and the like may turn out to be very helpful fictions, in much the same way that propositional logic is a quite interesting, yet highly fictional logic. The point is that epistemic logic is really worth looking into. Thus I am also claiming that the project to formulate a theory of mathematical practice will benefit from the use of formal tools such as epistemic logic. In the best of cases, it should be possible to formulate theorems about the nature of mathematical practice. As must be obvious, this position is *not* similar to Wittgenstein's attitude. Without going into details, one example may suffice to make the distinction clear. For Wittgenstein, the social coherence of the mathematical community does not need to be explained. It just happens to be that way, and it would not make sense to ask a mathematician why he or she is willing to accept the verdict of his or her colleagues as final.[9] In this case, I want to find arguments that explain the (necessity of the) coherence. One argument mentioned surely is that if A wants to check the proof of B, it increases efficiency if A and B share the same mathematical style. But the latter feature is precisely an important element that contributes to social coherence. It is, at the same time, a refusal to let the history, psychology, sociology, and economy of mathematics degenerate into a loose collection of interesting, anecdotal (therefore accidental) bits and pieces. The second reason, related to the first one, is that it is still possible to maintain the existence of a unique mathematical universe while holding the view that the way mathematics is done is best described using an RIM or RSM type of model. What is said here will sound only too familiar to any philosopher. The only thing that I am claiming is that the minimal realist position holds for mathematics as well. It does not follow—and I emphasize this point most strongly—that taking a sociological, psychological, or whatever point of view implies the impossibility of the existence of something like *the* mathematics. True, one might argue that a separate entity such as the unique mathematical universe is not called for, but, as must be clear, an appeal to the practice of mathematics to deny its existence does not carry the force many authors expect or want it to do.

It would be wishful thinking to believe that the above plea will bring together Type I and Type II philosophers of mathematics. Perhaps they do not need to be brought together physically. If a common language is available—and a modest proposal for a candidate is sketched out in this chapter—it will be there for anyone who wants to use it. Now, all too often, a false dichotomy is drawn.

NOTES

A first draft of this chapter was presented at the Center for Philosophy of Science, University of Pittsburgh, October 1988 on the invitation of Jerry Massey. This version has benefited from criticisms both from Jerry Massey, Ken Manders, and the other fellows of the center present at the time. Ken Manders's criticisms were especially important, but, taken seriously (as they should be), they constituted a new research program.

1. An excellent overview of the literature is provided in Thomas Tymoczko, 1986. A few additional important works not mentioned in this chapter are: David Bloor, 1976; Sal Restivo, 1983; and Eric Livingston, 1986 (for a critical review of Livingston, see David Bloor, 1987). See also Philip Kitcher, 1988.

2. See also Douglas Bridges and Fred Richman, 1987, for an excellent and introductory discussion and presentation of principles such as the Limited Principle of Omniscience.

3. For the epistemic mathematics approach, see Stewart Shapiro, 1985. Related articles have been written by Nicolas D. Goodman, 1984; and Stewart Shapiro, 1980. A historically interesting contribution is Kurt Gödel, 1969.

4. Classics in the field of epistemic logic are Jaakko Hintikka, 1974, 1975; and Karel Lambert, 1969.

5. Besides the example of Euler's conjecture mentioned in the text, errors have occurred in Fermat's Last Theorem (see Van Bendegem, 1987), in the Four Color Theorem (Ian Stewart [1987:111] speaks of a "comedy of errors"), and in the Goldbach conjecture, the Riemann hypothesis, and the Bieberbach conjecture.

6. See Carl B. Boyer, 1959. Wallis's proof dates from 1656–57 and is to be found in his *Opera Mathematica* (Boyer, 1959:168–74).

7. It would be a mistake to believe that the practice of confining inconsistencies is typical for the mathematical period preceding the age of rigor. Dirac's delta-function is a similar, quite recent case.

8. The first version is to be found in Alan Baker, 1984:4; the second version is taken from Daniel Shanks, 1978:6–7. The corollary of theorem 6 states that "If a prime p divides a product of numbers, it must divide at least one of them."

9. This short excursion into the Wittgensteinian field is not meant to take into account all the intricate details and complexities one finds in Wright (1980) and Shanker (1987).

REFERENCES

Arruda, A. I.
1980 "A Survey of Paraconsistent Logic." Pp. 1–41 in A. I. Arruda, R. Chuaqui, and N. C. A. da Costa (eds.), *Mathematical Logic in Latin America*. Amsterdam: North-Holland.

Baker, A.
1984 *A Concise Introduction to the Theory of Numbers*. Cambridge: Cambridge University Press.

Beeson, M.
1985 *Foundations of Constructive Mathematics*. Heidelberg: Springer.

Bishop, E.
1967 *Foundations of Constructive Analysis*. New York: McGraw-Hill.
1985 "Schizophrenia in Contemporary Mathematics." Pp. 1–32 in M. Rosenblatt (ed.), *Errett Bishop: Reflections on Him and His Research*. Providence, R.I.: AMS.

Bloor, D.
1976 *Knowledge and Social Imagery*. London: Routledge and Kegan Paul.
1987 "The Living Foundations of Mathematics." *Social Studies of Science* 17, 2:337–58.

Boyer, C.
1959 *The History of the Calculus and its Conceptual Development*. New York: Dover.

Bridges, D., and F. Richman
1987 *Varieties of Constructive Mathematics*. Cambridge: Cambridge University Press.

Ginsberg, M. L., ed.
1987 *Readings in Nonmonotonic Reasoning*. Los Altos: Morgan Kaufmann.

Gödel, K.
1969 "An Interpretation of the Intuitionistic Sentential Logic." Pp. 128–29 in Jaakko Hintikka (ed.), *The Philosophy of Mathematics*. Oxford: Oxford University Press.

Goodman, N. D.
1984 "The Knowing Mathematician." *Synthese* 60, 1:21–38.

Gorenstein, D.
1986 "Classifying the Finite Simple Groups." *Bulletin of the American Mathematical Society*, 14:1–98.

Hintikka, J.
1974 *Knowledge and the Known: Historical Perspectives in Epistemology*. Dordrecht: D. Reidel.
1975 *The Intentions of Intentionality and Other New Models for Modalities*. Dordrecht: D. Reidel.

Kitcher, P., guest editor.
1988 "Philosophie des Mathématiques (Philosophy of Mathematics)", special issue of *Revue International de Philosophie* 42, 4.

Lakatos, I.
1976 *Proofs and Refutations: The Logic of Mathematical Discovery*, ed. J. Worrall and E. Zahar, Cambridge: Cambridge University Press.

Lambert, K., ed.
1969 *The Logical Way of Doing Things*. New Haven: Yale University Press.

Livingston, E.
1986 *The Ethnomethodological Foundations of Mathematics*. London: Routledge and Kegan Paul.

Priest, G., R. Routley, and J. Norman, eds.
1989 *Paraconsistent Logic: Essays on the Inconsistent*. Munich: Philosophia Verlag.

Restivo, S.
1983 *The Social Relations of Physics, Mysticism, and Mathematics*. Dordrecht: Reidel.

Shanker, S. G.
1987 *Wittgenstein and the Turning-Point in the Philosophy of Mathematics*. London: Croom Helm.

Shanks, D.
1978 *Solved and Unsolved Problems in Number Theory*. New York: Chelsea.

Shapiro, S., ed.
1985 *Intensional Mathematics*. Amsterdam: North-Holland.

Shapiro, S.
1980 "On the Notion of Effectiveness." *History and Philosophy of Logic* 1:209–30.

Steiner, M.
1975 *Mathematical Knowledge*. Ithaca: Cornell University Press.

Stewart, I.
1987 *The Problems of Mathematics*. Oxford: Oxford University Press.

Tymoczko, T., ed.
1986 *New Directions in the Philosophy of Mathematics*. Stuttgart, Boston: Birkhauser.

Van Bendegem, J. P.
1985 "Dialogue Logic and Complexity." Gent, Belgium Manuscript.
1987 "Fermat's Last Theorem seen as an Exercise in *Evolutionary Epistemology*." Pp. 337–63 in W. Callebaut and R. Pinxten (eds.), *Evolutionary Epistemology: A Multiparadigm Program*. Dordrecht: D. Reidel.
1988 "Non-Formal Properties of Real Mathematical Proofs." Pp. 249–54 in A. Fine and J. Leplin (eds.), *Philosophy of Science Association 1988*, vol. 1. East Lansing: PSA.

Versluys, J.
1914 *96 bewijzen voor het theorema van Pythagoras* (96 Proofs of Pythagoras' Theorem). Amsterdam: A. Versluys.

Welti, E.
1987a *Die Philosophie des strikten Finitismus: Entwicklungstheoretische und mathematische Untersuchungen über Unendlichkeitsbegriffe in Ideengeschichte und heutiger Mathematik*. Bern: Peter Lang.
1987b "The Philosophy of Strict Finitism." *Theoria* 2, 5–6;575–82.

Wright, C.
1980 *Wittgenstein on the Foundations of Mathematics*. London: Duckworth.

3

A Naturalized Epistemology
for a Platonist Mathematical Ontology

Numbers, sets, functions, and other paradigmatic mathematical objects are, according to the Platonist view, outside spacetime and incapable of interacting with ordinary bodies within it. Taking them thus provides a nicely satisfying metaphysics of mathematics, but it appears to create an immense epistemological gulf between us and the mathematical realm. It is therefore hard to see how we can encompass mathematical objects within our most compelling model of the acquisition of knowledge, a perceptual model, where physical interactions play a central role. For centuries this apparent epistemological contrast between mathematical and physical entities has motivated empiricist critiques of mathematical Platonism.

Paul Benacerraf threw out the empiricist challenge for a generation of philosophers of mathematics in 1973 (Benacerraf, 1973) when he required a satisfactory account of mathematical knowledge to be a species of a general causal epistemology. Since then we have found that causal theories of knowledge stumble over even ordinary material bodies (Maddy, 1982). Yet, Benacerraf's demand was based upon good empiricist intuition: Any satisfactory epistemology should explain our knowledge of mathematical objects without endowing them or ourselves with occult properties or faculties. In today's epistemological circles, this demand often translates as an insistence that the epistemology of mathematics be naturalized.

In this chapter I will take some first steps toward meeting the challenge to naturalize the epistemology of mathematics. I do not do this merely to try to cover myself with a modish mantle, though sticking with fashion at least guarantees one partners in philosophical dialogue. Rather, I do this because meeting the empiricist challenge, in whatever form it currently assumes, is the dialectically strongest position for me as a Platonist to take.

During most of this chapter I will be setting the stage for a postulational account of the genesis of mathematical knowledge. My hypothesis is that our mathematical ancestors brought mathematical objects within our cognizance by positing them. This suggestion raises many questions concerning how positing can generate knowledge about preexisting entities—especially how it can do this when the entities are mathematical ones. At the end of the chapter, I will hint at answers to such questions. The bulk of the chapter will be concerned, however, with addressing the question of how a postulational account of mathematical knowledge could count as a piece of naturalized epistemology—even if it is successful in its own right.

The processes referred to in a naturalized account of knowledge must be natural processes. I will argue later that positing is such a process. Yet, we can posit supernatural objects, for example, spirits, as well as natural ones. Thus positing, no matter how natural a process, can never lead to knowledge of mathematical objects, if they are not natural objects themselves. Now, Quine defines naturalism so that it counts mathematical objects as natural objects, but David Armstrong defines naturalism so that it excludes mathematical objects from the natural universe. To avoid begging the question against Armstrong, I will begin by arguing for a place for mathematical objects within the naturalist's ontology. Then, I will try to specify general parameters for a naturalized epistemology for mathematics, and sketch my postulational account of the origins of mathematical knowledge. Finally, I will prescind some of the problems positing mathematical objects purports to pose.

I will assume the truth of Platonism throughout the chapter, and I shall not argue for it directly. (Of course, the chapter argues for it indirectly, since it tries to disarm one of the major objections to Platonism.) Furthermore, the reasons, described below, that might have led our ancestors to posit mathematical objects probably remain good reasons—although certainly not the only reasons—for us to accept mathematical objects today.

MAKING ROOM FOR MATHEMATICAL OBJECTS

Can a naturalist countenance mathematical objects? Unless the answer to this question is affirmative, our quest for a naturalized epistemology for mathematical objects is bound to fail. So we must address this question before proceeding further.

Yet before we can do that we must answer still another question: What is naturalism? Despite the currency the naturalist philosophy enjoys, I was surprised to find that there is little consensus concerning its content. Quine writes as follows:

> Now how is such robust realism to be reconciled with what we have just been through? The answer is naturalism: the recognition that it is within science itself, and not in some prior philosophy, that reality is properly to be identified and described. (Quine, 1981)

Here Quine throws ontological and epistemological questions into the court of science, leaving open the possibility that Platonic objects are real. And, as is well known, Quine does think that science ontically commits us to such mathematical objects.

David Armstrong's characterization of naturalism is both more definitive and importantly different from Quine's:

> Naturalism I define as the doctrine that reality consists of nothing but a single all-embracing spatio-temporal system. (Armstrong, 1981:149)

Armstrong's position seems to rule out Platonic mathematical objects at the start—a consequence Armstrong is quick to acknowledge.

For Philip Kitcher, writing in the philosophy of mathematics, naturalism goes without a definition, although he declares himself a naturalist and allies himself with empiricism against a priorism (Kitcher, 1988). Evidently, we must decide on a characterization of naturalism before we proceed much further.

Plainly, our target should be Armstrong, since he is completely candid about having no place for Platonic mathematical objects within the "single all-embracing spatio-temporal system" that defines his naturalist universe. Since mathematical entities have no effect in that system, "there is no compelling reason to postulate them" (Armstrong, 1981:154).

It is striking that Armstrong excludes mathematical objects from the naturalist's ontology, while Quine admits them. Armstrong is aware, of course, that Quine and other philosophers argue that we must postulate mathematical objects in order to do science and that our justification for doing so is no different from that used to justify positing electrons and other theoretical entities (Armstrong, 1981:155). But that cuts no ice with him. Consider his forceful response:

> There is this vital difference. [Classes, and so forth] provide objects the existence of which, perhaps, can serve as truth-conditions for the propositions of mathematics. But this semantic function is the only function they perform. They do not bring about anything physical in the way that genes and electrons do. In what way, then, can they help to explain the behavior of physical things? Physics requires mathematics. That is not in dispute. But must it not be possible to give an explanation of the truth-conditions of mathematical statements in terms of the physical phenomena that they apply to? (Armstrong, 1981:155)

The issue here, then, is not that these objects have been introduced as posits but rather that they have no causal powers and play no role in explaining the behavior of physical things. Of course, if it is essential to naturalism that its objects have causal effects on things in the physical world, then Armstrong wins his case hands down. But he also suggests that naturalists can recognize objects that "can help to explain the behavior of physical things." This may be just the crack in his argument Platonists need.

Armstrong concedes that mathematics plays an essential role in physics but denies that its objects play an explanatory role. What function, then, do these objects have? There is the semantic function Armstrong already mentioned: we cannot use mathematics in calculations and deductions unless its terms refer to and its sentences have truth-values.[1] But mathematics is more than a device for calculating and reasoning about physical phenomena; it also helps us describe them. Mathematical ideas infect virtually all of physics. Even at the elementary level, concepts such as instantaneous velocity (change in displacement at an instant, ds/dt) and momentum (mass X velocity) defy reformulation in nonmathematical terms.[2] At the more advanced levels of science we find phase spaces, vector and tensor fields in physics, growth functions and probability distributions in biology, and demand curves and utility spaces in economics. Without such mathematical entities these sciences could not even begin to describe the phenomena they recognize today. Now one cannot explain physical phenomena that one cannot describe. So certainly mathematical objects "can help to explain the behavior of physical things," at least in the sense of being an indispensable tool for the task.

What is more, mathematical facts and properties of mathematical objects play essential roles in physical explanations themselves. Consider this explanation of why a ball thrown straight up in the air reaches a specific point (rather than another) before it comes back down.

At any instant the velocity of the ball (i.e., the speed and direction with which it travels) is the resultant (vector sum) of its upward and downward velocities. When thrown it has an initial upward velocity of, say, v^* and a zero downward velocity. However, the force of gravity subjects the ball to a positive downward acceleration, a^*, which in turn increases its downward velocity over time until the latter equals and then exceeds its upward velocity. When the two velocities are equal the ball stops its upward course. The downward velocity is identical to a^*t, so the ball stops its upward flight when and only when

$$v^* = a^*t.$$

Thus it will stop when and only when $t = v^*/a^*$. But this value of t determines the exact upward displacement of the ball.

Plainly, this explanation appeals to several mathematical properties of the ball's velocity, for instance, that it is the vector sum of its upward and downward velocities and that when these are identical the velocity equals zero. Moreover, this velocity itself, being a function, is a mathematical object. So the explanation uses mathematical objects and their properties to explain the behavior of a physical thing.

Such considerations do much to undermine Armstrong's argument; but I can easily imagine his defenders protesting that they do not count, because I still have not found any causal role for mathematical objects. This brings us back to Armstrong's primary reason for excluding mathematical objects—their lack of causal powers.

Rather than dealing with this head on, I shall examine a more general assumption implicit not only in Armstrong's thinking but also in many writings in contemporary philosophy of mathematics. This is the assumption that a clear and sharp, causally or spatiotemporally grounded, ontic division obtains between mathematical and physical objects.

The assumption probably comes from thinking about the relative scope of logic, mathematics, and physics: Logic properly includes mathematics; mathematics properly includes physics. Hence it is tempting to see sharp lines between logic and its ontology (or lack thereof), between mathematics and its ontology, and physics and its ontology. It is tempting to think that we can excise the mathematics from physics in order to achieve a fully naturalized ontology.

In finding no mathematical objects within the "all-embracing spatio-temporal system," Armstrong may be presupposing a space-time criterion for differentiating between (abstract) mathematical and (concrete) physical objects: the latter but not the former are within spacetime. But what is it to be in spacetime? To be located in it? To be part of it? To be either? Are spacetime points in spacetime? Is all of spacetime in itself? These are not idle questions. The ontic status of the universal gravitational and electromagnetic fields, prima facie physical entities, as well as that of spacetime points, prima facie mathematical entities, turns on how we answer them. Furthermore, each answer comes with its own set of unresolved controversies (cf. Hale, 1988; Resnik, 1985b).

Moreover, even quantum particles, such as electrons, widely regarded as paradigm physical objects, pose difficulties for a locationally grounded division between the mathematical and the physical. Where are these particles when they are not interacting with each other? On one interpretation of quantum theory, under some circumstances, these particles are not even located within a finite region of spacetime. Then, are they everywhere or nowhere? The answers are controversial, and so is the interpretation of quantum theory, but until the issue is settled we can hardly be satisfied with classifying entities by means of a spacetime criterion.

Quantum particles seem more like mathematical objects than like everyday, commonsense bodies. To see why, read this excerpt from a recent text on particle physics:

> In the most sophisticated form of quantum theory, all entities are described by fields. Just as the photon is most obviously a manifestation of the electromagnetic field, so too is an electron taken to be a manifestation of an electron field and a proton of a proton field. Once we have learned to accept the idea of an electron wavefront extending throughout space . . . it is not too great a leap to the idea of an electron field extending throughout space. Any one individual electron wavefront may be thought of as a particular frequency excitation of the field and may be localized to a greater or lesser extent dependent on its interactions. (Dodd, 1984:27)

But what is a field? The simplest precise description is that it is a function defined on every point in space whose value at that point gives the intensity of the field there. Add to that the reflection that in quantum mechanics talk of intensity at a point (or in a region) is really talk of the probability of an interaction taking place there, and you see how mathematical is the quantum field theoretic conception of particles.

Another commonly accepted way of distinguishing between the physical and the mathematical is to claim that mathematical objects cannot change properties and participate in events. But this way will not stand up to a first set of objections either. Numbers do change some of their properties. Numbering the wives of Henry VIII was only a fleeting property of the number one. Perhaps such properties should not count, on the grounds that they are just accidental properties of numbers. But physical objects can change only their accidental properties too. Perhaps numbering the wives of Henry VIII should not count because it is not a real property. But what makes it unreal? Surely not because it is a relational property, since electrons, say, have properties only by virtue of their relations to other particles. I do not take these quick jabs to be fatal blows to the idea that numbers cannot change, while physical objects can, but they do show that some careful work must be done before we can use the idea of change to distinguish mathematical from physical objects.

Similar difficulties surround the event side of the proposal. The functions used to characterize physical events change their values during the course of an event. We have already seen that in the example of the thrown ball. It is question begging to simply object at this point that the velocity function does not participate in such an event because it cannot. Better to say that the event can be fully and precisely described without referring to the functions. But can such a redescription be carried out? I suppose that in the ball case it might, although the difficulties attending Hartry Field's nominalization should give us pause. However, we have no reason to think that it can be done with events involving subatomic particles, whose basic features, such as charge, spin, and energy level, correspond to no commonsense ideas. This should also remind us that fields participate in events too; they collapse and interact. But fields seem to be hybrid entities hovering between the paradigmatically mathematical and the paradigmatically physical.

The considerations also suggest that it is unwise to exclude objects from the naturalist's universe on the grounds that they have no causal powers or that they play no causal role in explaining the behavior of physical things. For we have seen that it can be unclear whether a given explanatory object (for example, a field) is physical or mathematical or even whether something counts as physical behavior (for example, the collapse of a field).

The tendency for physicists to seek structural explanations of the fundamental features of physical reality also undermines the idea that a fundamental ontic division obtains between the physical and mathematical. The movement began with Einstein's identification of the gravitational field with spacetime itself, which in turn identified masses and their gravitational effects with variations in the geometric structure of spacetime. More recently, physicists have proposed that all of physical reality is an eleven-dimensional space, whose geometrical properties give rise to all of the known physical forces (Freedman and van Nieuwenhuizen, 1985). The trend has been, then, to pass from conceiving of subatomic particles as tiny bodies to conceiving of them as systems of interacting fields spread over spacetime and thence to local variations in the structure of a generalized spacetime. From the point of view of today's science, physical reality is most accurately described as an unchanging structure, whose local variations may be described in less sophisticated terms as bodies and causes, changes and happenings. How then can naturalists recognize just bodies and causes, changes and happenings as real?

Perhaps they will declare that the space—however complicated—embraced by physics is real and then add that other "purely mathematical" spaces are unreal. But what can this mean? It cannot mean that physical reality instantiates the structure in question, for, on the line we have been following, physical reality is nothing but that very structure. A much more appealing answer is to introduce the idea of an observable as a type of local geometric variation and then argue that physical space is real, because in it alone are all and only possible observations also observable events. On this reading, the difference between physical space and other spaces is not that the latter do not contain observable events—for some will, since they are just local structural features—but rather that other spaces differ from physical space by failing to contain all and only the "events" humans could observe. This is an epistemic difference, an important one to be sure, but not enough of one to distinguish physical space ontically from other purely mathematical spaces. None of the difficulties we have encountered with distinguishing physical objects from mathematical objects need arise, if we adopt Quine's characterization of naturalism and let science tell us what exists. Nor need we worry about the existence of mathematical objects, since in asserting that particles have velocities, that reactions reach equilibrium points, and so forth, science commits itself to them again and again.[3] So I will answer in the affirmative the question with which this section began—naturalists can countenance mathematical objects. We are now free to seek a naturalist epistemology for them.

Before we do so, let me deal with the impression one might form that I have been inconsistent in denying the ontic distinction between mathematical objects and physical objects. After all, I began this chapter by emphasizing the abstractness of mathematical objects and the apparent epistemic gap between them and ordinary physical bodies, and lately I have been arguing that the distinction between the abstract and concrete blurs in theoretical science. But I have not denied that there appears to be a striking epistemic gap between ordinary bodies and mathematical objects. Platonists must struggle with the epistemology of mathematical objects, so long as the relative abstractness of mathematical objects seems to prevent the epistemology of ordinary bodies from applying to them. I should also add that Platonism does not need a sharp cleavage between the abstract and concrete for its metaphysics of mathematics to work. Platonism succeeds because, unlike nominalism, materialism, and constructivism, it can supply the vast infinities of objects that mathematics requires—more objects than any mind or minds could construct, more objects than the physical universe contains.[4] Whether these objects be fundamentally different from material or mental ones is not crucial. (For convenience I will continue to refer to mathematical objects as abstract and to ordinary physical bodies as concrete.)

WHAT IS NATURALIZED EPISTEMOLOGY?

As in the case of naturalism qua metaphysical doctrine, there is a perplexing variety of opinions concerning the definition of naturalized epistemology. Hilary Kornblith introduces his anthology on the subject by stating that, in contrast to traditional epistemology, naturalized epistemology holds that the answer to the question "How ought we arrive at our beliefs?" is not independent of the answer to the question "How do we arrive at our beliefs?" and recognizes that psychological investigations can be directly relevant to epistemological ones (Kornblith, 1985). Kornblith's characterization includes normative work, such as Alvin Goldman's, within the scope of naturalized epistemology.

Quine, the source of the term *naturalized epistemology*, formulates his idea of naturalized epistemology in this passage: "Epistemology, or something like it, simply falls into place as a chapter of psychology, and hence of natural science." (Quine, 1985:24). The difference between knowledge and mere true belief is usually taken to be normative: knowledge is true belief that passes epistemic muster. To the extent that psychology is not concerned with norms, epistemic or otherwise, Quine's suggestion that epistemology become a branch of psychology appears to leave behind an important part of the theory of knowledge.[5]

Now I think that Quine's approach is not as far from Kornblith's as the foregoing passages indicate. By letting epistemologists use methods from social sciences other than psychology, we can keep their enterprise within the spirit of Quine's view while also permitting them to describe our epistemic norms and account for the evolution of these norms.[6] Epistemologists can also go beyond natural history and evolutionary theory, if we allow them to systematize our epistemic norms using the method of reflective equilibrium applied in logic, linguistics, and descriptive ethical theory (Resnik, 1985a). As the case of logic shows, organizing our norms thus might yield insights into them more valuable than the system itself.

However, describing, systematizing, and explaining our epistemic practices is one thing; evaluating them is something else—apparently beyond the scope of science, social or physical or formal. Yet even a fair amount of evaluation and criticism could be brought within the purview of epistemology qua science, if such evaluation and criticism were approached from the applied scientific viewpoint of an efficiency engineer. Given a description of our epistemic values and a measure of desirable performance (efficiency), epistemic engineers could determine how close our actual epistemic practices come to the official standards, according to the official yardstick. They could even reform our practices by suggesting methods for bringing our performances closer to the received standards. That is as far as we can go towards normative epistemology without allowing epistemologists to make their own value judgments. That may be far enough. And we may be at the limits of naturalism.

On the other hand, naturalistic epistemologists are members of the scientific community and, as such, free to promote new epistemic values from within that community. In doing so they act no longer as naturalistic epistemologists per se, just as political scientists put their academic roles aside when stepping inside the voting booth. Such actions are consistent with the naturalist's credo, so long as our epistemologists forsake supernatural normative insights.

If the preceding thoughts correctly represent the spirit of Quine's approach, there is no real dispute between him and Kornblith over the limits of naturalized epistemology. Even if it is real, we need not settle it now, since the bulk of the account I will present here is genetic rather than normative.

That is not to say that no work awaits normative epistemologists in the philosophy of mathematics. For all the successes of mathematical logic, we still do not have a good understanding of why we insist on proof in mathematics, why we prize alternative proofs, or of how axioms are justified—to name just a few of the questions that come to mind.[7]

Our concern here, the genetic side of epistemology, is especially pressing for the Platonist. Winning a place for Platonic mathematical objects within the naturalist's universe does not even begin to address the question of how to give a naturalized account of the genesis of our knowledge and beliefs about them. To deal with this problem, we must first develop a firmer conception of what a naturalized account of cognition is.

Cognition is a process, hence a naturalized portrait of it will present it as a natural process. Following Armstrong's definition of naturalism, one could define a natural process as one that takes place wholly within spacetime. This would exclude any theory that involved mathematical objects themselves in the process of cognizing them. As it is, I will not offer such a theory, so I could be at home with an Armstrong-style characterization of natural processes. Yet, on the grounds of consistency and liberality, I suggest the we follow Quine and let a natural process be one that science "has identified and described."

One evident problem with this definition is that the limits of science are vague and unclear. Are psychoanalysis or intensional semantics part of science? If not, then naturalistic epistemologists may not explain the genesis of knowledge in terms of processes hypothesized by those theories. Perhaps they can avoid the problem of demarcating science by not straying into the fuzzy areas between hard-core science and the more controversial disciplines vying for scientific status.

Due to the immature state of the cognitive and social sciences, that may be easier said than done. Many seemingly natural human processes—communicating, learning from experience or from other people, developing preferences or creating theories and works of art—are so complicated that science has only the roughest understanding of them. Undoubtedly hard-core science will eventually bring them under its umbrella in some, perhaps unforeseeable, form. Yet, naturalized epistemologists may need to appeal to these processes now.

Well, let them. But let us restrict naturalistic epistemologists to processes appropriate to the study at hand. Obviously, one should not appeal to communication in giving a naturalized account of communication, but one should be able to appeal to our ability to recognize sound patterns. In a similar way, I think that it will be proper for me to appeal to our ability to acquire knowledge and beliefs about everyday bodies in accounting for the genesis of mathematical knowledge. Of course, someone might point out that we still do not have a satisfactory naturalized account of our knowledge of ordinary bodies. I would be among the first to agree. My idea is to let epistemologists build speculative theories on somewhat less speculative foundations. If we can show how our ability to know and refer to ordinary bodies can by natural means generate an ability to know and refer to mathematical objects, then we will have made the prospects for a naturalized epistemology for the latter all the greater. And we will have shown that if we cannot have natural knowledge of mathematical objects, then it is unlikely that we can have it of ordinary bodies.

I am proposing to give naturalistic epistemologists liberty to hypothesize processes that we might not be able to manipulate experimentally, due to our current lack in knowledge and technology. Furthermore, they should be free to appeal to (hypothetical) events in the remote history of our species. Let us recognize, however, that if they do make such an appeal, then we may never be able to put their theories to a direct experimental test. Such accounts are already common in evolutionary biology and other historical sciences.[8] Evidence for and against them can be sought and is often found. So while requiring naturalized epistemologists to suggest experimental means for testing their theories is too stringent, it is reasonable to ask that they suggest some connection between their speculations and observable evidence.

Testing theories about the cognition of mathematical objects also seems to raise special problems, for we cannot manipulate mathematical objects experimentally. Thus we cannot study human cognition of them as we can human perception of ordinary bodies, where we can alter states of subjects, bodies, and even media in order to see how each affects reports by the subjects of what they perceive. But this difficulty arises only for those who think that some mathematical knowledge is acquired by something akin to perceiving mathematical objects. I disavow this sort of approach.

In these last paragraphs I have urged what I take to be a moderate approach to naturalized epistemology. My approach falls on a spectrum that ranges from the conservative to the speculative. The most conservative approaches recognize only processes that are clearly and uncontroversially described by natural science and that are also subject to experimental control. That would mean describing cognition in terms drawn from physics, chemistry, anatomy, and parts of biology. I know of no epistemologist so rigorous. At the speculative end of the spectrum we find people, for example, Penelope Maddy, willing to posit new cognitive processes or faculties, such as an ability to perceive certain sets of concrete objects via their members. Clearly this process is not subject to experimental manipulation, since we cannot place an informational screen between the set and its members to allow a subject to see one without seeing the other.

MATHEMATICAL KNOWLEDGE
AS KNOWLEDGE ABOUT PATTERNS

In my version of Platonism (Resnik, 1981), mathematical objects are positions in patterns, and mathematical knowledge is knowledge about patterns. On the one hand, this knowledge incompasses much more than the ability to recognize and distinguish simple patterns. Pigeons can do as much. On the other hand, it does not involve some sort of direct perception of patterns qua abstract entities, not even of those patterns that we "see" in perceptible arrangements. Knowing a pattern, in my view, is like knowing a theory. To know a theory is to know what entities and processes it posits and the behavior its laws attribute to both. Similarly, knowing a pattern is a matter of knowing what positions it contains and how they are related to each other.

In discussing knowledge about patterns it is important to distinguish the question of how a research mathematician learns about patterns from the question of how a learner in our society does, and this in turn from the question of how ancient humans learned about them. The mathematician has a gigantic collection of techniques the latter two lack.

I will focus on the question of how ancient peoples might have come to know patterns. One reason for doing this is that I am trying to answer skepticism concerning our ability to acquire knowledge about objects as abstract as my patterns. Focusing on the contemporary mathematician would be to ask instead how we manage to learn about new abstract entities once we already have an abundant fund of knowledge about some abstract entities. So I want to consider people who have no mathematical knowledge and suggest a natural process through which they could acquire it.

I thought about concentrating on the acquisition of mathematics by children. But our children have help from those already in the know. Furthermore, although it is possible that they recapitulate the process through which the human race learned mathematics, the rapidity with which they do so hampers studying it. I will also be concerned with the global question of how humans came to recognize and countenance mathematical objects as a kind of thing rather than with more local questions about how they came to recognize particular mathematical objects such as zero or the square root of two.

Since nobody knows how we developed mathematics, my story is perforce purely hypothetical. Despite this, it is easy to think of the kind of evidence that could bear upon it. Unfortunately, obtaining the evidence itself is much harder. Anthropological studies of primitive peoples could support or correct the initial elements of my narrative, but then a huge evidential gap opens due to the lack of cultures intermediate between us and still existing primitive peoples. History is of little help, too. Elsewhere I argued that the transition from premathematical studies (without ontic commitment to mathematical objects) to full-fledged mathematics (with ontic commitment to them) occurred when Babylonian and Eygptian mathematics developed into Greek mathematics (Resnik, 1982). We know something about the initial stage of this transition from fragments found by archeologists, and we know much about the final stage through Euclid and later commentators. To my knowledge, however, we have no useful evidence about the critical intermediate periods.

With the stage for my account now set, let us try to imagine ourselves in the situation of a primitive people who have no mathematics. Our knowledge of patterns will begin, like our knowledge of everything else, with experience. Experience will also teach us that certain shapes and arrangements work better in certain situations than others. Things having various shapes or arranged in certain patterns will become important to us.

Because of their practical importance, we will find ourselves driven to invent a vocabulary to name some of these patterns. The time will come, however, when we need to instruct a foreigner or novice, and nothing ready to hand is of the right shape. Then we might use a drawing. If so, we will have taken an important step: because we will no longer be restricted to indicating, recognizing, or labelling present things of the same pattern, we will be able to represent how things are arranged or shaped without having those things present.

We need some technical terminology at this point lest we confuse patterns in the concrete sense—in the sense in which we have a drawer full of dress patterns at home—with patterns qua abstract entities, for example, a dress pattern no one has described or drawn. A paper-and-ink dress pattern is an instance of an abstract pattern, for it is a token, a concrete inscription, of a symbol type. However, it is also a concrete representation of a type of dress without being an instance (token) of that type—without being a dress. Right now my concern is with the use of concrete inscriptions to represent other concrete things.

Reserving the term *pattern* for abstract entities, I will use the term *template* to refer to our usual concrete devices for representing how things are shaped, structured, or designed. Concrete drawings, models, blueprints, and musical scores are my paradigm everyday templates. Under the appropriate conventions, templates represent other concrete things, such as buildings, artifacts, or performances, which fit them in the appropriate ways. Templates are thus templates for things of the appropriate kind: blueprints are designs for buildings rather than for sculptures or performances of ballets.

We will go quite some distance in our practical talk about how things are shaped, arranged, or designed without appealing to abstract entities. Concrete templates will do the job perfectly. We will also learn how to design things before we start to manufacture them and how to use modified designs to learn things about modifications in the things themselves. Without introducing abstract entities, we will now be in a position to talk about possibilities, about how things might be arranged or designed or shaped.

Templates have two dimensions. Syntactically they are configurations constructed according to certain conventions. Semantically they represent other concrete things by means of implicit and explicit rules of representation. So far we have only considered templates that successfully fill their representational role. But we can also use the medium for constructing templates of a given kind to construct configurations without any representational role, such as random doodlings on blueprint paper. Surely, we will do that too.

We have thus advanced from the barest recognition of the practical importance of certain shapes and arrangements to a representational system for designing and thence to playful and creative attempts to explore possibilities. Before we move on, let us remember that we use language to construct templates too; for we can often describe an arrangement, shape, or design in words more accurately than we can in a drawing. Some linguistic templates will be sets of directions—instructions on how to build a serviceable lean-to, for instance. Others will describe rather than instruct—a biologist's description of bee dances is an example. Later in mathematics, linguistic templates will be our chief and most reliable methods for representing patterns.

I have not yet touched on the crucial question of how our experience with templates could lead us to knowledge of abstract patterns. The discussion of templates is not in vain, though, because it indicates how we might have begun our initial explorations of patterns and our initial probing of the possible. It also tells us something about the local epistemology of patterns. For although templates now only represent

concrete things, they will come to represent the abstract patterns concrete things fit. Looking at the example of a dress, we see that ultimately there will be four entities involved, two concrete ones—the
dress and its template, and two abstract ones—the dress pattern and
the symbol type of the template. These are related according to figure
3–1.

Thus once we take the step towards countenancing patterns qua
abstract entities, it is likely that we will see patterns as associated with

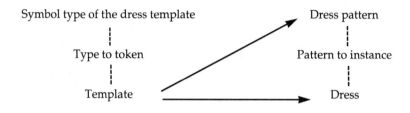

Figure 3–1.
Paradigm for Pattern Representation.

templates according to figure 3–1. This will indicate to us that we can
construct and study templates to gain information about patterns. Finally, having bonded patterns and templates, it will become plain that
to show that a specific pattern for, say, houses exists, it suffices to exhibit a blueprint for houses of that pattern.

Another important point to notice is that we could not have got-
ten this far with templates without developing complex syntactic systems (such as place notations for the numbers and geometric diagrams), as well as conventions and rules through which these systems
represent concrete entities and criteria for determining whether a
given configuration counts as a coherent representation. All of these
contribute to easing the step towards full-blown mathematical theories with genuine commitments to mathematical entities.

We might have been led to introduce abstract entities in order to
make sense of unending progressions. Many ordinary phenomena
might have prompted us to wonder about them, but we can count on
our (by then) well-developed numerical notations and geometrical
templates to have led us to think of counting without end or subdividing lines into smaller and smaller segments. Perhaps we jumped to abstract mathematical objects at this point. But we might have resisted
the move. Good nominalists among us might have shown us how to

account for our intuitions (about counting endlessly and ceaselessly subdividing lines) in terms of the possibility of performing more and more actions—actions which would require, of course, ever more matter, ever increasing lifetimes and attention spans, ever diminishing marks, and the like.

Thinking of mathematical objects as possible concrete ones hits its limit when it comes to limit entities. Perhaps stretching a cord tighter and tighter eventually forces it to be perfectly straight, but subdividing a line into smaller and smaller segments cannot yield an extensionless point; and drawing finer and finer lines cannot produce lines without breadth. Thus in countenancing limit entities it will no longer make sense for us to speak in terms of possible concreta. We will be forced to posit entities, such as points, lines, and circles, as sui generis and existing in their own right or else forego them altogether.

It would have also been simpler and more perspicuous for us to construe limit entities as abstract instead of trying to make do with possibilities. Even today it is unclear how limit entities could be formed from ordinary concrete material by natural processes. The only suggestion that I know of would be to see each limit entity as arising through the completion of an infinite process. One might think, for example, that it is possible to divide a region infinitely many times until nothing but an unextended point remains. But the history of mathematical attempts to understand infinite sequences and sums (as well as Zeno's paradoxes and their variants) shows that talk of completing an infinite process is a metaphor at best, one that breaks down when we ask what counts as finishing the process, what the last steps are like and, what results they produce. Thus it is more intelligible to deny that limit entities are some sort of actual or possible concreta and to posit them as nonmaterial and timeless things to which our concrete objects at most approximate. This move would deftly forestall questions concerning the origins and material properties of limit entities.

Yet such a move would have prompted skeptics to demand an explanation of how one can acquire knowledge about entities so different from the objects of our ordinary experience. I have brought our ancestors to the brink of recognizing abstract entities. Their colleagues' point is that it is not rational to take the plunge unless we (that is, our ancestors) can provide an account of how we can acquire knowledge about the new entities once we have countenanced them.

One way to start would be to posit, in addition to patterns, isomorphisms between certain simple finite templates and the patterns associated with them. Using these we could project some properties of templates onto patterns. We would find that much of our knowledge of the former transferred to the latter.[9] Furthermore, some of the discoveries about templates that led us to limit entities would also help us to discern some of the latter's properties. The considerations that showed us why circles, points, and lines cannot be identified with concrete geometric inscriptions would also indicate where such inscriptions reliably reflect features of abstract entities and where they fail. Thus, we would not be asking our fellows to accept a new mystery that we alone are qualified to interpret. We could point out that we and they have already clearly discerned some of the features of these new entities, that we already have a scheme for representing many of their features, and that we already have some methods for determining their properties and have reason to expect to develop more. Finally, we could also point out that if we do take the plunge and countenance limit entities, then we could also countenance numbers, linguistic types, and a host of other abstract entities our templates represent more adequately than they represent limit entities.

It would have been unfair and premature, however, for our ancestors' skeptical colleagues to demand a complete account of the methods for learning about abstract patterns. Our ancestors would have hardly begun to see the ontological picture. They might have reasonably expected to discover then-undreamed-of methods for exploring the new ontology, just as they could not have conceived of some of the ways we now have for learning about objects as familiar to us as our own bodies.

POSITING MATHEMATICAL OBJECTS

Many abilities developed by humans prior to the onset of mathematics figure in the account of the last section, including the ability to communicate, to use pictures, diagrams, and words to represent things that are absent or merely imagined; the ability to speculate; and, finally, the ability to hypothesize and theorize about new kinds of entities. According to some philosophers, many of these abilities already require interacting with abstract entities. Frege, for instance, maintained that speaking a language, judging, reasoning, indeed, thinking of any kind took place through "grasping" thoughts—his term for abstract entities associated with sentences as their meanings. If he and others like him are right, then my account presupposes an ability to interact with abstract entities at its outset. If they are right, the entire project of naturalizing epistemology seems doomed from the start. More brightly, if they are right, then, of course, the question of how we know mathematical objects is not fundamentally different from the question of how we know anything about anything at all.

I do not think that Frege was right nor do most contemporary philosophers of cognition. None of them claim to understand the mechanisms involved in thinking, representing, and communicating significantly better than Frege did, but they judge his approach to be a scientific dead end. (One reason is that a system of sentences and brain states can take over the role thoughts play in Frege's account of cognition and communication.) Thus, under the current circumstances, it is fair for us to assume that no abstract entities participate in the premathematical activities that eventuate, on my account, in mathematical knowledge.

Positing mathematical objects first brought them under our ken. Although positing mathematical objects is a variety of verbal behavior, it does not involve a Fregean grasping of abstract entities. To posit mathematical objects is simply to introduce discourse about them and to affirm their existence. It involves nothing more mysterious than the ability to tell fairy stories, invent myths about the gods, or theorize about the forces at work in the observable world. Even Armstrong implicitly recognized as much by questioning our justification for positing mathematical objects without casting aspersions on our ability to do so.

Although I have spoken of positing so far in connection with introducing a whole new category of objects, mathematicians and scientists also use it to introduce single objects within an extant framework. We see physicists positing new particles as additions to extant systems, astronomers affirming new galaxies, and mathematicians postulating new transfinite cardinals. Also, the line between positing and discovering often blurs. Thus people speak of the discovery of the positron, although Dirac posited it long before any one elicited its observable traces. For our purposes, however, it is unimportant that these lines fade.

Yet here is a puzzle we should face, if only briefly. People have posited ghosts, the Ether, and phlogiston with as much ease as they have posited numbers. How can positing lead to knowledge in the one case and not in the others? What distinguishes between them? Primarily, truth and existence. Ghosts, the Ether, and phlogiston do not exist. Hypotheses that they do, are false. Hence no matter how justified people might have been in positing them, doing this could not have led them to knowledge. Of course, truth and existence are no guarantee that positing will lead to knowledge, since positors may lack the appropriate justification for their true beliefs. But our ancestors did not lack an appropriate justification for believing that mathematical objects exist, and they do exist (or so I have assumed throughout this chapter); so our ancestors' positing led them to mathematical knowledge.

As to ourselves, most of us acquire our initial beliefs about mathematical objects from teachers. On the face of it, then, we acquired our mathematical knowledge by having it communicated to us by those in the know. This is no more problematic or unnaturalizable than our ability to learn from our teachers and texts about historical figures or foreign lands and peoples. If a naturalized epistemology can make sense of the transmission of knowledge in the one case, it should be able to make sense of it in the other.

A more subtle problem concerns the aboutness of our mathematical beliefs. What makes them about mathematical objects? And in what sense are they about them? Are they about the same objects that our mathematical ancestors posited? A related problem concerns the apparent lack of "epistemic contact" with mathematical objects which positing does not seem to provide. Some might think that this means that we cannot have knowledge of mathematical objects. This worry is probably due to a mistaken adherence to the mathematics/physics distinction. I think that I can put it to rest along with the other worries I have just canvassed. Unfortunately, I cannot even begin to do so here.[10]

NOTES

I would like to thank Dorit Bar-On, Michael Hand, William Lycan, and Susan Williams for their help in writing the paper this chapter is based on.

1. This point, originally due to Frege (cf. Resnik, 1980:62–63), was emphasized by Benacerraf (1973). Circumventing it was the principal motivation for Field (1980).

2. Field's failed attempt (1980) underscores both the conceptual and technical difficulties of such a project. I discuss the former in Resnik, 1985b and both sorts of problems in Resnik, 1985c.

3. In speaking thus I have not been thinking of science as encompassing so-called pure mathematics. It thus falls short of affirming the existence of many of the entities studied by the far reaches of contemporary mathematics. I do not find this a drawback at this point, since my purposes in this paper will be served if I can naturalize the epistemology of the kind of mathematical entities that figure in science. Furthermore, I think that it is arguable, along the lines I have been pursuing so far, that pure mathematics is part of science.

4. Even the totality of all spacetime points and regions falls short of the number of objects required by mathematics.

5. It is not clear that Quine meant to exclude normative investigations, since he writes that we are still prompted to do epistemology for the traditional reasons, "namely, in order to see how evidence relates to theory, and in what ways one's theory of nature transcends any available evidence" (1985:24).

6. Quine may have restricted himself to psychology on the grounds that the other social sciences make such heavy use of the problematic ideas of translation and interpretation that they fail to count as genuine sciences. But even psychology trades heavily in beliefs, which are under Quinean interdiction along with translation. So since none of the social sciences can be taken over intact, properly filtered treatises on (at least) anthropology, history, and sociology should lie alongside of psychology in the epistemologist's library.

7. Important work on these questions is contained in Kitcher, 1983 and Maddy, 1988a, 1988b.

8. For a recent study of such explanations see Resnik (forthcoming a, b).

9. I discuss this in more detail in Resnik, forthcoming a.

10. I address these questions in Resnik, forthcoming b.

REFERENCES

Armstrong, D.
1981 "Naturalism, Materialism and First Philosophy." In D. M. Armstrong, *The Nature of Mind and Other Essays*. Ithaca: Cornell University Press.

Benacerraf, P.
1973 "Mathematical Truth." *Journal of Philosophy* 70:661–80.

Dodd, J. E.
1984 *The Ideas of Particle Physics*. New York: Cambridge University Press.

Field, H.
1980 *Science without Numbers*. Princeton: Princeton University Press.

Freedman, D. Z., and P. van Nieuwenhuizen
1985 "The Hidden Dimensions of Spacetime." *Scientific American* 252:74–81.

Hale, S. C.
1988 "Spacetime and the Abstract/Concrete Distinction." *Philosophical Studies* 53:85–102.

60   *Philosophical Perspectives*

Kitcher, P.

1983 *The Nature of Mathematical Knowledge.* New York: Oxford University Press.
1988 "Mathematical Naturalism." In P. Kitcher and W. Aspray (eds.), *History and Philosophy of Modern Mathematics,* Minnesota Studies in the Philosophy of Science, 11. Minneapolis: University of Minnesota Press.

Kornblith, H.
1985 "Introduction: What is Naturalistic Epistemology?." In H. Kornblith (ed.), *Naturalizing Epistemology.* Cambridge, Mass.: Bradford Books, MIT Press.

Maddy, P.
1982 "Mathematical Epistemology: What is the Question?" *Noûs* 16:106–7.
1988a "Believing the Axioms." Pt. 1. *The Journal of Symbolic Logic* 53:481–511.
1988b "Believing the Axioms." Pt. 2. *The Journal of Symbolic Logic* 53:736–64.

Quine, W. V. O.
1981 "Reply to Stroud." In *Foundations of Analytic Philosophy,* Midwest Studies in Philosophy, 6. Minneapolis: University of Minnesota Press.
1985 "Epistemology Naturalized." In H. Kornblith (1985).

Resnik, D. B.
1989 "Adaptationist Explanations." *Studies in the History and Philosophy of Science* 20, 2:193–213.

Resnik, M. D.
1980 *Frege and the Philosophy of Mathematics.* Ithaca: Cornell University Press, 1980.
1981 "Mathematics as a Science of Patterns: Ontology and Reference." *Noûs* 15:529–50.
1982 "Mathematics as a Science of Patterns: Epistemology." *Noûs* 16:95–105.
1985a "Logic: Normative or Descriptive? The Ethics of Belief or a Branch of Psychology?" *Philosophy of Science* 52:221–38.
1985b "How Nominalist is Hartry Field's Nominalism?" *Philosophical Studies* 47:163–81.
1985c "Ontology and Logic: Remarks on Hartry Field's Anti-platonist Philosophy of Mathematics." *History and Philosophy of Logic* 6:191–209.
1992 "Proof as a Source of Truth." In M. Detlefsen (ed.), *Proof and Knowledge in Mathematics.* London: Routledge and Kegan Paul.
1990 "Beliefs About Mathematical Objects." In A. Irvine (ed.), *Physicalism in Mathematics.* Dordrecht: Kluwer Academic Publishers.

4

Mathematical Skepticism: Are We Brains in a Countable Vat?

By the end of this chapter, I will propose a rather extreme socio-logical view of a branch of mathematics. I will contend that the development of transfinite set theory is guided not by some external domain of sets or by some transcendental logic but by basic features of human nature. However, I develop my suggestion rather slowly. In particular, I will try to develop for mathematics a version of the traditional philosophical problem of skepticism about the external world. My mathematical skepticism will be directed at the distinction that lies at the core of modern set theory, the distinction between countable and uncountable sets. I want to know whether this distinction has any objective reality.

Allow me to begin by setting the issue in a context.

In the anthology *New Directions*, I espoused the cause of quasi-empiricism as an approach in the philosophy of mathematics, borrowing the term from Lakatos and Putnam (Tymoczko, 1986). There are two key aspects to quasi-empiricism: first and foremost, an emphasis on mathematical practice; and second, an openness to scientific methods in mathematics, for example, the use of computer proofs. Although these two features are compatible, sometimes they can pull in opposite directions.

The second feature, which invites us to see mathematical methods as close to scientific methods—or closer than was once thought—goes naturally with a realist view of mathematical objects and with the image of mathematicians as discovering the nature of mathematical reality (just as scientists are imagined to discover the nature of physical reality). This view is most natural with regard to computer proofs like that of the Four Color Theorem (Tymoczko, 1986) and the recent proof that there is no finite projective plane of order 10 (Browne:1988). In such cases, it is easy to believe that there is a mathematical object, a "formal proof" (actually a mathematical pattern of a certain sort), whose existence we become aware of by empirical means. As one of the authors of the latter proof, Clement Lam, said:

> in this kind of problem, the mathematician can not personally check each step of a complete proof. But with computers, mathematical proofs are becoming more like the experiments conducted in the physical sciences than the traditional proofs of Euclidean geometry. When you get a result with a computer, the best you may be able to do is to show that the same result will be obtained if someone else repeats the experiment in just the same way. (Browne, C17:1988)

Moreover, this scientific strand of quasi-empiricism can be extended to other branches of mathematics in the spirit of Quinean holism (Tymoczko, forthcoming). We could posit real numbers, functions, and even sets as part of our overall scientific theory of the world. But this assimilation of mathematics and science makes most sense in the area of applied mathematics, where we posit, for example, the objects of Newtonian calculus to enable us to express Newtonian physics. The case is altogether different when we turn to pure mathematics.

In a recent article, Penelope Maddy (1988) attempts to make the case for quasi-empirical methods in evaluating potential set theoretic axioms such as the Continuum Hypothesis or the existence of Measurable Cardinals. Maddy's idea is that just as scientific statements are accepted on probabilistic evidence that falls short of conclusive corroboration, so too might set theoretic axioms be accepted on the basis of merely probabilistic arguments. In my opinion, her argument is technically flawed, since the quasi-empirical methods she discusses are all seriously inconclusive: probabilistic arguments for a given axiom are balanced by probabilistic arguments against it. That is, she presents examples of quasi-empirical evidence that might favor set theoretic statement T only when there is equally good quasi-empirical evidence that favors statement not-T. Thus, there is really no analogy to science where probabilistic arguments actually do carry the day, because the evidence for one statement, as opposed to its negation, can predominate.

Furthermore, Maddy's quasi-scientific picture of set theory makes sense only if we adopt a strong realist stance toward the existence of a determinate universe of sets, that is, if we presuppose that there is a complete and definite totality of sets independent of human experience. It differs from classical realism, or Platonism, only by allowing that our intuitions of this mathematical reality might not be exhausted by traditional proofs but can be supplemented by probabilistic arguments and inductive evidence. It is at this point that the tension with the emphasis on mathematical practice becomes most apparent.

In the *New Directions* anthology, I identified quasi-empiricism in the philosophy of mathematics as that approach that begins with and seeks to understand the actual practice of mathematicians. One motive for this was to free philosophy from a priori conceptions of what mathematics must be like—especially the conception that mathematical results must be a priori, but also from other foundational programs. This emphasis on mathematical practice is naturally seen as an empirical or sociological approach to mathematics, an approach that René Thom characterizes, with some justice, as the view that "a proof P is accepted as rigorous if it obtains the endoresement of the leading specialists of the time" (1986:71). This view of mathematics as a human practice or a cultural institution is championed by mathematicians such as Wilder, Davis, and Hersh, philosophers, especially Wittgenstein, also Lakatos, Bloor, Restivo, myself, and others such as De Millo, Lipton, and Perlis. The more one focuses on the particularities of the mathematical community and its practices, the less one focuses on an alleged reality that they are uncovering. Wittgenstein (1956, 1976) was especially adamant in his attacks on realism or Platonism in mathematics and in his insistence that mathematics was essentially different from science.

Thus the first moral I draw is that quasi-empiricism, like politics, makes strange bedfellows. Many people can claim, or be claimed under, the banner of quasi-empiricism, but the differences among us can be quite dramatic. In this chapter, I will make a tentative stab at sorting out some of these differences. I will focus attention on one branch of mathematical practice, contemporary set theory, but I will treat it as a branch of mathematics like any other, ignoring claims that set theory is a foundation of mathematics. Does this practice commit us, or its practioners, to a belief in an independent universe of sets, and, if so, what further commitments does that belief entail? My strategy will be to recreate in set theory a version of the classical skeptical dilemma in epistemology. In the end, I will hazard some conclusions about the "sociological view" of mathematics, but the value of the chapter such as it is, is in the connections it makes along the way to those tentative conclusions.

Skepticism is a central topic of modern philosophy. Yet although skepticism poses a formidable challenge to empirical knowledge and to knowledge of other minds, one does not often read of mathematical skepticism as a challenge to mathematical knowledge. There are several reasons for this apparent lack.

In the first place, skepticism turns on a distinction between appearance and reality. The evidence for our beliefs is derived from how things appear to us and the truth of our beliefs is determined by how things actually are. Skepticism begins by noting the gap between the two, but in the case of mathematics, it seems that somehow appearance and reality are one and the same. A dreamt proof is still a proof: "Whether I am awake or asleep, two and three add up to five, and a square has only four sides; and it seems impossible for such truths to fall under a suspicion of being false" (Descartes, in Anscombe and Geach, 1969:63).

In the second place, by attacking mathematical knowledge, the skeptic is in danger of biting off more than he can chew. Mathematics is so close to rationality, to the very possibility of logical discourse, that if it is called into question, the framework of questions and answers is undermined. It is true that Descartes went on to ask "may not God likewise make me go wrong whenever I add two and three, or count the sides of a square, or do any simpler thing that might be imagined?" but he wisely does not pursue the point (Descartes, in Anscombe and Geach, 1969:64). For how could we meet this question by rational argument? No matter how convincing a reply we could construct—let if have the full rigor of a mathematical proof—the skeptic could counter that for all we know it only seems convincing to us and in reality we are "going wrong." Philosophical skepticism, as opposed to blind irrationalism, presupposes we are rational, and that is pretty close to presupposing mathematics. Without this, the skeptic's game is over before it has even begun.

A third reason for the apparent absence of mathematical skepticism might be that it is not really absent at all but goes by a different name. On one reading of traditional skepticism, it is not so much empirical knowledge that comes into question but the interpretation of that knowledge: our claims to knowledge of a physical world existing independently of our perceptions. An analogous version of mathematical skepticism would question not mathematical knowledge but our claims to know of a mathematical reality existing independently of our proofs. Thus construed, mathematical skepticism turns out to be anti-Platonism or anti-realism, the denial of a mathematical reality independent of us. Thus, both formalism and intuitionism might be seen as versions of mathematical skepticism.

Nevertheless I am unhappy with the way this analogy leaves things. If one is, as I am, dissatisfied with both formalism and intuitionism for the many cogent reasons given in the literature, one seems to be driven back to Platonism, and the issue of skepticism is swept under the rug. More precisely, many Platonists (such as Maddy and Gödel) attempt to avoid skepticism by postulating a faculty of mathematical intuition that is supposed to insure our direct contact with the mathematical realm.[1] However, when the matter is put in the context of skepticism, the appeal to intuition seems to beg the question. No one who engaged with the traditional skeptic about our knowledge of the external world or our knowledge of other minds would accept as a solution that we just intuit the external world or other minds—as if we needed only to introduce a word to defeat skepticism!

Let me return a moment to the first two reasons given above for the lack of discussion of mathematical skepticism. These were our apparent inability to distinguish between appearance and reality in mathematics and the essential connection between mathematics and rationality. I want to suggest that these reasons do not apply with equal force to all of mathematics. I concede their validity for some branches of mathematics, such as number theory. But I will contest their validity for contemporary transfinite set theory. However plausible it is that first-order logic and number theory constitute the notion of rationality, it is not at all plausible that the Axiom of Choice, the Continuum Hypothesis, or even Replacement is essential to rationality. Moreover, by developing a version of Skolem's Paradox, I'll attempt to draw the contrast between appearance and reality in mathematics—specifically by suggesting that for all we know we might really be living in a countable world. This way of putting "the Paradox," besides being colorful, severely tests some purported resolutions including appeals to mathematical practice. After examining some of these, I'll present a solution based on Putnam's resolution of the traditional brains-in-a-vat dilemma and conclude with some general comments about quasi-empiricism.

Traditional skepticism, at least one form of it, begins with the distinction between appearance and reality. More exactly, it makes the claim that what justifies our ordinary beliefs must be internal processes, experiences, and ratiocinations, while what makes these beliefs true is their correspondence to features of external reality. The skeptic then goes on to argue that our inner experiences and thoughts could remain exactly the same, even though the outer world were radically different—we could, for example, be dreaming or deceived by an evil demon. In a modern version, the skeptical possibility is expressed in the parable of brains in a vat. It is consistent with physical principles that the universe could have evolved (or thrown together) a giant computer connected to brains kept in a vat of nutrient fluid in such a way that the computer feeds the brains just the elaborate systematic experiences we actually have! Thus the hypothetical brains would have the same experiences we do, even though their environment and ours are incredibly different. If this is possible, it seems that even the totality of available evidence cannot tell between the vat alternative and the usual more comfortable picture of an enduring world of physical objects.

This form of skepticism has a serious point: it is that our overall picture of the world with ourselves in it does not hang together well. There is, in principle, a great gap between our evidence and what we take ourselves to claim. So the skeptic's challenge to us is not merely to solve a puzzle but to deal with the gap in our conceptual system that has apparently been uncovered.

It is worth noting that skepticism does not depend on demonology or even science fiction. The gap can be brought out by taking our scientific picture of the world very seriously. Quine does this by starting with our best scientific picture of the world and us as physical beings in it. He then draws a line around each of us at the limits of our sensory receptors. Any information we can get about the outside world, he claims, must be carried by energy transfers across this line. But, Quine notes, this evidence is very meager and does not uniquely determine a theory of the external world; in fact, the totality of all possible evidence radically underdetermines scientific theory (1960, esp. Chap. 1; 1969).[2] We cannot, as it were, get back to where we started by reasoning from the evidence. Now the brains-in-a-vat scenario is just one way to make this predicament graphic.

Given this way of looking at things, there are two natural ways around skepticism (three, if we count Descartes's heroic attempt—he calls upon an omnipotent and all-loving God who insures that our proper justifications ["clear and distinct ideas"] correspond with reality). One path is to stick to justification and inner experience and to abandon external reality. So phenomenalists try to reconstruct our apparent talk of the external world as talk about sense data. The other path is to stick to reality and truth and to abandon justification. So materialists (in a current manifestation, reliabilists), argue that we're just bodies in a physical world and if our senses are reliable indicators of our environment, then that is all we can ask.

If mathematical skepticism is a cogent possibility—and that is yet to be established—we can draw some parallels with the above. Formalism (conventionalism, constructivism) is the obvious analogue of phenomenalism, giving up truth and reconstructing reality in terms of justification. But Platonism is a very poor analogue to materialism; it is hard to draw causal connections between the unchanging world of mathematical reality and our occasioned beliefs. Rather, Platonism is the analogue of heroic Cartesianism with mathematical intuition playing the role Descartes assigned to God. With this much as a background, let us attempt to reconstruct the skeptic's argument inside mathematics.

In his famous Diagonal Argument, Cantor proved that the real numbers are uncountable (in the process, he gave mathematical significance to the distinction between countable and uncountable sets).[3] Later, Skolem showed that any consistent theory, including the theory that Cantor used for his proof, could be interpreted over a countable model. Thus a mathematician living in such a countable model could use Cantor's very argument to prove that the reals in his or her world were uncountable. From outside the model, however, we could see that the reals in it, indeed the entire model, form a countable set. This appears to be a paradox: it appears that the reals are both countable and uncountable. This appearance is sometimes called "Skolem's Paradox."

But if it is a paradox, it is not a very serious one and can be dissolved by observing that a referential shift occurs when we pass from one model of a language to another (where "language" is interpreted as a purely syntactical object). Because the countable world is a model for set theory, mathematicians in it can prove the same theorems of set theory that we can. Hence they can prove the sentence "The reals are uncountable." But in their usage "the reals" refers not to what we refer to by the same word, but to a set in their universe. We know, and can prove by our construction, that that set is countable. However, by "uncountable," they refer to the concept of uncountablity in their model. They give the same formal *definition* of uncountable as we do—there does not exist a one-to-one function from the natural numbers onto the set in question. But this definition picks out a different property in their world, just as the word *reals* picks out a different set.

Their reals are uncountable in their world because there is no appropriate one-to-one counting function in their world. Thus the apparent contradiction is dissolved. The reals really are uncountable. What they call "the reals" in their model really are not the reals but a countable subset. However they have no notion of a reality beyond their model (that's what it means to have a model or interpretation of their language). So what they mean by the sentence "The reals are uncountable" is true in their world.

Having resolved one paradox, however, we find another problem arising to take its place. How do we know that we are not living in a countable model of set theory? The mere fact that we can prove that the reals are uncountable in our terms no longer seems sufficient, since mathematicians in a countable model could apparently give the same proof in their terms! Does it make sense to suppose we might be living in a countable universe? Prima facie, yes, as long as we take the strong Platonist stance that our mathematics is about a reality independent of humans. In just the way that it makes sense to assume we might be brains in a regular vat, hooked up to an independent reality in a perverse way, it makes sense to assume we might be in a countable world, hooked up to mathematical reality in a perverse way. After all, couldn't God just erase all but countably many mathematical objects from existence without our noticing the difference? Or, if this puts too much strain on God's power, suppose the computer tinkered with the brains in a vat just a bit so that their faculty of mathematical intuition was altered. Whenever they attempted to grasp the reals (or the universe of sets) they succeeded only in grasping a countable subset (or a countable submodel). They would be brains in a countable vat.

So it seems, at the moment, that there is a real possibility that we might be brains in a countable vat, as real a possibility as that we might be brains in the more usual sort of vat. Would that matter? Some people are inclined to say no, just as they would to the standard skeptical possibilities. After all, if we can't tell the difference, why should it matter? But this is to ignore the conceptual challenge posed by skepticism. If we could be living in a countable world, then that possibility seems to undermine the essential distinction between countable and uncountable sets (just as the possibility of living in a usual vat world undermines the concept of physical object). As Skolem came to believe, the possibility could be interpreted as showing that the distinction is relative, not absolute.[4] Uncountable sets might not be *big* sets, just *complicated* ones. What would be threatened is our way of picturing mathematics, what Wittgenstein called mathematicians' "prose." It is not that we couldn't go on to do set theory in the same way, given the skeptical possibility, but that once we recognized it, we might not want to go on as we had previously. We might come to regard talk of large cardinals and other things as excessively metaphorical, as puffed up. Mathematicians might come to believe, as Wittgenstein hoped they would, that Cantor's Paradise was really no paradise at all and so leave of their own accord (1976:103).

Before examining some solutions to Skolem's skeptical puzzle, let me comment briefly on this case of mathematical skepticism. In the first place it nicely parallels empirical skepticism. As we saw earlier, it is possible to argue that a thoroughgoing scientific picture of the world leads to a skeptical position—the evidence that we have, according to science, is insufficient to justify the science we begin with. So too, Skolem's skepticism can be seen as starting from a thoroughgoing set theoretic picture of the world of the sort Maddy endorses and leading to a skeptical position—the proofs that we have, according to mathematics, are insufficient to justify the mathematics we begin with.

In the second place, notice that it is precisely a realist conception of mathematical entities that gets us into trouble. This is par for the course: as we saw in the beginning of this chapter, skepticism trades on the distinction between what makes propositions true (relation to reality) and what justifies them (appearances—in this case, proofs). Ordinarily one explains mathematics by assuming there is a real world of sets and that somehow or other (God's grace) we latch onto the sets in the right sort of way. But now we notice that if we latched onto only a small subset of that totality, the countable sets, in the "right sort of way," then everything would go on as before. We cannot tell the difference between the naive Platonic connection and the perverse Platonic connection. Thus the idea that we are brains in a countable vat seems quite cogent, a very apt metaphor for the above predicament. And we have kept the promise of distinguishing between appearance and reality in mathematics. The appearance of an uncountable reality seems quite compatible with an actually countable reality.

Lastly, I suggest that although realism gives rise to skepticism, there is no way around skepticism from a realist position. Appeals to mathematical intuition just won't help, for the simple reason that mathematical intuition, albeit a subtly different sort of intuition, is operative for the brains in the vat. What distinguishes us from the brains is not the existence of an extra-mathematical reality (there is that in the mathematical vat), or a mysterious intuition connecting us to that reality (there is that, too, in the vat), or our ability to do proofs (the brains can create the same symbol strings we do). All these essential ingredients we share with the brains in a countable vat. In order to defeat skepticism, the realist needs some further hypothesis, like a Cartesian God, who insures our intuitions get connected up in the right way. Skolem skepticism is not easily resolved.

Let us now briefly review Skolem's two solutions to his paradox as they are discussed by Paul Benacerraf (1985). Originally Skolem presented his results as showing the inadequacy of axiomatic set theory as a foundation of mathematics. However, although the foundations controversy was raging in the early twentieth century, it is no longer very relevant today. Fortunately, we can excise foundationalist concerns from Skolem's first answer rather easily to get the conclusion that formalization, that is, axiomatization in first-order logic, is insufficient to capture informal mathematics (in particular, the theory of transfinite sets). Benacerraf seems to endorse this answer. The Skolem-Benacerraf view is that pure formalization admits unintended models, so that the intended model cannot be captured by this method.

This answer is quite familiar as one of the standard antiformalist critiques based on various limitation theorems. (Notice, by the way, that it presupposes that Skolem's results do show something prima facie untoward about mathematics. If those results were unexceptional—in our terms, if there were no possibility that we lived in a countable world—then nothing would be shown about formalism or anything else.) Notice further, that this answer turns on a particular realist conception of formalisms, that is, axiom systems are meant to characterize models and the failure of categoricity is the untoward result (this feature is not an objection to my development of skepticism, since that development, like the scientific development of traditional skepticism, begins by assuming the concepts and distinctions it will later call into question). Finally, the Skolem-Benacerraf answer will be completed only if we can explain exactly what it is that the formalist axiomatization leaves out. However, as I argued above, that explanation cannot be provided by general talk about a universe of sets and mathematicians' intuitions about that universe, for that general talk can be accommodated by the skeptic's example of brains in a countable vat. Moreover, it is hard to see how Benacerraf's appeal to "informal practice" can solve anything. Exactly what is it about our informal practice that guarantees that the reals are uncountable in a way in which a formal version of Cantor's proof apparently does not? According to the story, the brains appear to do exactly what we can do.

Skolem's second answer, according to Benacerraf, is the one he arrived at later in life and the one usually associated with his name. Skolem Two has accepted the formal method as providing all that can be said about mathematics. Consequently, he concluded that such concepts as 'countable' and 'uncountable' are essentially relative. Exactly what Skolem meant by "relative" is open to question, but the most natural explanation is the one alluded to above: there is no absolute distinction; rather, relative to an initial universe, countable and uncountable divide up the sets in that universe. Metaphorically, we might say that the concepts are relative in the sense that the brains-in-a-countable-vat scenario makes dramatic—their distinction is every bit as good as ours. In other words, there's really no telling whether or not we live in a "really countable universe." Indeed, if we take an extreme formalist position where the axiom systems are sufficient unto themselves (they are not attempts to characterize models of any sort), then the term *relativism* drops away as irrelevant, and all we are left with is a systematic and empty formal theory. The content of the countable-uncountable distinction reduces to nothing more than is given in the formal axioms. In particular, the view that uncountable sets are very large collections is nothing but a picture; there are no things outside the formal system itself to be large or small. (Here again we have an echo of Wittgenstein's attack on transfinite set theory: the extension from finite sets to infinite sets is totally unwarranted and only a colorful, but misguided, way of speaking suggests otherwise.)

In his reply to Benacerraf, Crispin Wright offers the following intriguing possibility for preserving the countable-uncountable distinction from the skeptic's threat.

I believe we can glimpse the possibility of an alternative to the platonism which, pending disclosure of some internal flaw in the skeptical arguments, was all that seemed available as an alternative to the skeptical conclusion . . . What we need to win through to, I suggest, is a perspective from which we may both repudiate any suggestion of the platonic transcendence of meaning over use and recognise that meaning cannot be determined to within uniqueness if the sole determinants are rational methodology and an as-large-as-you-like pool of data about use. Wittgenstein wanted to suggest that the missing parameter, the source of determinacy, is human nature. Coming to understand an expression is not and cannot be a matter of arriving at a uniquely rational solution to the problem of interpreting witnessed use of it—a "best explanation" of the data. Still less is it a matter of getting

into some form of direct intellectual contact with a platonic concept, or whatever. It is a matter of acquiring the capacity to participate in a practice, or set of practices, in which the use of that
expression is a component. And the capacity to acquire this capacity is something with which we are endowed not just by our
rational faculties but to which elements in our sub-rational natures also contribute: certain natural propensities we have to up-
hold particular patterns of judgement and response. (Benacerraf,
1985:117–37)

But it is not at all clear how to interpret this Wittgensteinian picture for the case at hand. Let me digress, for a moment, to give an example of when this picture would make sense: the Gödel case of unprovable formulas. One likes to think of the concept of number as determinate, so that each closed formula of arithmetic or its negation is
true. Now Gödel seems to have shown that if this is so, no formalization can capture the intuitive conception of number for any consistent
formalization is incomplete. Furthermore, the formalist's natural response, that there is intrinsic undecidability here, seems forced—we
are not undecided about Gödel's formula; we firmly believe "this sentence is unprovable" to be true. On the other hand, the Platonist's appeal to an independent mathematical reality grounding our intuitions
brings its own difficulties.

In this case, Wittgenstein seems to offer a genuine alternative in
just the manner that Wright suggests. Both formalism and realism are
wrong. But formalism is not wrong because it has failed to articulate
beliefs (intuitions) about the numbers we possess implicitly. It is not
that we somehow already possess the appropriate belief. Rather it is
that we are all inclined to go on in a genuinely new way. It is because
we are the kind of creatures that we are, because of human nature,
that we respond in the same way to Gödel's result, that we all accept
the unprovable formula as true and its unprovable negation as false.
We come to count something as a proof despite the fact it cannot be
represented in our formalism, but, in so doing, we are not being faith-
ful to some determinate mathematical reality independent of us, nor
are we relying on some intuition the formalist forgot to formalize. We
are going on in a new way, and it happens to be the same way for all
of us because we agree in "our sub-rational natures." Those who respond differently risk expulsion from the community of mathematicians.

Right or wrong, this picture has a certain plausibility when applied to Gödel, but how is it supposed to apply in Skolem's case?

There doesn't seem room for a choice here, no formula about which we can disagree. I don't see where our practice, informal mathematics, or subrational natures can take hold, can guarantee the countable-uncountable distinction is the one we really want and not some countable counterfeit. To be sure, I think there is something very significant in Benacerraf's and Wright's suggestions. These elements do contribute to our way of doing set theory and I will return to this later. But they cannot solve the dilemma of brains in a countable vat for the simple reason that all these elements can occur with the brains in a vat! If classical Platonism were true, it appears there would be no answer to mathematical skepticism.

There is an answer to the Skolem problem, and it is implicit in the work of Hilary Putnam (1981).[5] The marvelous thing about Putnam's argument is that it works equally well to show that we are not brains in a countable vat and to show that we are not brains in a vat of any sort! He is willing to grant, for the sake of argument, that there could be brains in a vat (countable or not). The thrust of the argument is that we are not they. Analogously, I might grant that there are possible worlds in which I don't exist, but still argue that ours is not such a world. One of Putnam's key insights is that the brains-in-a-vat situations are not deceived about anything. They aren't deceived, because their words don't mean the same as ours (if they did, at least one of us would have to be systematically deceived). But as we saw in describing the Skolem case, there is a referential shift that occurs when the same syntactic language is reinterpreted over the vat (countable or otherwise). What they mean, then, by the formula "the reals are uncountable" is something totally different from what we mean by the same formula. We have different thoughts, and that is why both can be true. Here is a summary of Putnam's argument:

1. We can raise such questions as are we brains in a vat? are there elm trees? are the reals uncountable? These are all legitimate questions in our language. We can raise them simply because of our mastery of our language.
2. But no brain in a vat could raise such questions. The reason for this is that, according to Putnam, most words have their meaning solely in virtue of referring to what they do. Meanings are not mental or brain states: "meanings just ain't in the head," as he puts it. So to ask a question about elm trees or the real numbers, one must be able to refer to them. But this is just what the brains in a vat can't do. When constructing that scenario, we left the elm trees and most of the reals out of the vat! If they ask

meaningful questions, it must be in reference to items in their vat universe.

3.   It follows, therefore, that we are not brains in a vat, countable or otherwise.

Putnam's argument about the general brains in a vat case strikes many people as suspect. I have defended it elsewhere, so I won't repeat that defense here. What I would like to emphasize is how much more plausible it is when deployed against the Skolem Paradox. After all, there is something crazy about the suggestion that we are living in a countable model. We can prove that the reals are uncountable. That proof gives meaning to that term. Ultimately—and this is Putnam's deep point—it doesn't make sense to wonder whether the reals are "really countable" according to some hypothetical, alien concept of "countability" to which we, by hypothesis, have no access. We are stuck with our words in the universe we inhabit. The very idea that we are in a countable universe is semantic nonsense, a grammatical illusion. For in that idea, the words *countable* and *universe* have been divorced from their usual mathematical meaning (according to which the question is easily answered "no"). The words are now supposed to get their meaning somewhere else, from the outsiders' point of view. But from our point of view, that question—Are we living in a countable universe? as part of an alien language—is no more meaningful than, Are we living in a "gffflh pssst"?

To put the point another way, the countable model exists as a possibility only relative to our world. It is defined to be different from our world, an interpretation alternative to the standard (that is, original) one we begin with. We construct the alternative by leaving certain things out of the model (for example, almost all the real numbers). It is totally incoherent to turn around and wonder whether *that* world might not be ours after all. Moreover, this argument is not one bit weakened by the fact that mathematicians in a countable world can give it, for it expresses a deep point about reality. The point is that the distinction between countable and uncountable sets is an internal distinction, a distinction within our mathematical theory. (What else could it be?) It makes no sense to wonder whether sets really are countable in some absolute sense, apart from the sense we, following Cantor, give this term in the context of our mathematics. (Exactly the same kinds of reasons can be given by Putnam when showing that we are not brains in an ordinary vat.)

Thus the best argument that we are not brains in a countable vat happens also to show that we are not brains in any kind of vat. I take the extreme implausibility of the mathematical skepticism to give credence to the more general argument. Many other philosophers have expressed doubts about the more general argument and so they are left with the problem of brains in a countable vat.

In both cases, Putnam's arguments turn on a philosophy of language, an account that stresses the role of use or practice in determining meaning. For Putnam, there are no mental meanings that mediate reference. Our words, like *3, pi, omega, reals*, achieve their semantic import in terms of the references they make. In this sense, Putnam's philosophy of language is very realist; reference is the basic semantic relation. But reference is not a magical or puffed up relation. In the end, it is explained by the simple schema that referring term *t* refers to t. To say that words essentially refer is not to posit some magical relation between words and a collection of objects that exist independently of us. Putnam calls his view "internal realism," for the reference relation is construed as an internal relation between two different aspects of our conceptual scheme, words and objects.

Thus Putnam's internal realism does not commit set theory to the unbridled realism that Maddy defended in the article mentioned in the beginning of this chapter. To say that there are sets, infinite, even uncountable sets, is just to say that some set theoretical terms are referential—even less, that such terms don't have mental meanings. It does not follow from this that there is a full and determinate set theoretic reality such that every proposed axiom is either true or false of it independently of our conceptions. Rather, in the ideal, what is true is simply what is rational for us to accept (Putnam, 1981:preface). (Recall Thom, "a proof is what convinces mathematicians.")

So the Putnam solution to the Skolem puzzle is, like the quasi-empiricism I defended in *New Directions*, neutral with respect to the degree of realism we espouse. We can be realists while acknowledging the intrinsic connection between reality and our own view of it. But what of all that talk of informal mathematics and human nature that appeared in Benacerraf, Wright, and (by implication) Wittgenstein? If it doesn't solve this version of the Skolem Paradox, then what is its point in discussing set theory?

Wittgenstein was quite adamant about what he thought were conceptual confusions that surrounded set theory. "Infinite sets are not large sets," he warned, and he claimed that set theory made the fundamental error of treating finite collections and infinite series (or laws) with the same notation, a notation for arbitrary sets.[6] The mistake, Wittgenstein said, was to read into the technical notions of set theory, one-to-one mappings, countability, uncountability, and so forth, analogies with considerations of size about finite domains. Indeed, Wright goes on in the aforementioned article to raise similar doubts. Perhaps the countable-uncountable distinction has nothing to do with size but only with the relative complexity of certain sets. (Notice how close this answer comes to the purely formalist reply. All there are are the technical mathematical notions which might as well be called C-able and unC-able. It is a mistake, according to Wittgenstein, to color these notions with shades borrowed from the finite domain. It is sullying mathematics with unnecessary and harmful "prose.") Notice, too, that at this point, Wittgenstein is interfering with the mathematicians in just the way that he claimed that philosophy had no right to do. "But I'm not interfering with their mathematics," he could claim, "just their prose. I don't want to drive them out of paradise, just show them set theory is not a paradise and then they'll leave of their own accord."

Most mathematicians, I feel sure, will reject these Wittgensteinian objections. Indeed, as Shanker makes clear (unintentionally), the objections are unconvincing and often reduce to a sort of moral outrage at the obstinacy of mathematicians (Shanker, 1987). But I propose a novel way of saving Wittgenstein from himself. Suppose that the grounding of our concepts, our human nature, or natural inclinations, what Wright calls "sub-rational" faculties and natural propensities, include making the very analogies that Wittgenstein inveighs against. That is, what if what Wittgenstein regarded as the prose, the color, is part of the mathematics and cannot be separated from it? We are creatures that are naturally inclined to see sets as including infinite and finite sets, to see infinite sets as large, and uncountable sets as especially large, all on the model of size relations among finite sets. This needs no justification, this is just how we act. So there is more to the concepts of set theory than is provided by any formal system. But that is provided more by our own natures than by any external reality. In other words, I'm suggesting that it matters what we *call* our mathematical concepts. The term *large cardinals* inspires and guides mathematical activity in a way that "curious and intricate set theoretic constructions" cannot. Mathematicians prove things in set theory by imagining the transfinite universe, by taking their prose seriously, just as science fiction writers solve their problems by taking their imaginary universes seriously. In neither case do we have anything beyond the practices and expert practitioners. In both cases the practices give meaning that no formal specification can exhaust, because the meaning includes the color, the direction, and the excitement of the practices.

If these last remarks are at all correct, then we have a novel account of Wittgenstein's own hostility to the infinite. He becomes a perfect example of someone drummed out of the mathematical community for failing to catch on! As his view allows and he himself admitted, there can be people who don't share the required inclinations of mathematicians—and Wittgenstein doesn't. He describes the way a tribe teaches its children how to count as follows:

> The children of the tribe learn the numerals in this way: They are taught the signs from 1 to 20 . . . and to count rows of beads of no more than 20 on being ordered, "Count these". When in counting the pupil arrives at the numeral 20, one makes a gesture suggestive of "Go on", upon which the child says (in most cases at any rate) "21" . . . If the child does not respond to the suggestive gesture, it is separated from the others and treated as a lunatic. (Wittgenstein, 1960:93)

When learning set theory, mathematicians are presented finite sets of increasing size and, finally, infinite sets and are told the latter are very much larger than the largest finite sets presented. Eventually they are shown Cantor's Diagonal Argument and told that this shows the reals are vastly larger than anything seen so far. They are encouraged to name other large sets. A mathematician who cannot do this, refuses to do this, complains about unwarranted transference from the finite domain to the infinite, is separated from the tribe of mathematicians and treated as a lunatic. Well, perhaps not as a lunatic, but certainly not as a mathematician. This is, I suggest, what happened to Wittgenstein, at least with regard to set theory. He could see alternatives where others could not; he could see what held mathematical practice together was neither insight into a Platonic realm nor complete specification of a formal system but nothing more than human nature, the shared inclinations, the natural propensities, the subrational nature of human mathematicians. But at a crucial point, his own inclinations led him astray. He is, to use his own term, a "lunatic" not because his views on infinite sets are crazy but because they are not shared by his community.

In conclusion then, it seems that mathematical practice, not scientific methodology, is what really matters to quasi-empiricist approaches, at least in pure set theory and perhaps the rest of pure mathematics. If we are to take a realist attitude at all, then only Putnam's form of internal realism can save us from Skolem skepticism. However, the realism that emerges does not justify a quasi-scientific attitude in set theory. There is no transfinite world out there, complete in all its details, waiting to be discovered. The countable-uncountable distinction is objective for the simple reason that the human community of mathematicians finds it reasonable to agree on this, and it will remain objective only so long as that agreement persists.

When we learn to accept the notion of an arbitrary set of integers, our acceptance is not grounded in a transfinite world; we are rather like readers who learn to accept the idea of time travel. In both cases, acceptance amounts only to continuing on with the story. The realism of set theory is a feature of mathematical practice, of the way we talk about sets, of the metaphors we use and the analogies we draw, of our mathematical prose and the proofs we choose to puff up. Wittgenstein was right to notice the puffed up proofs and the colorful prose of set theoreticians. He was wrong to think that set theory, or any mathematics, can be done without such color. Set theory is not, as Maddy suggests, a quasi-science. It is a quasi-art.

NOTES

1. If one is moved to Platonism for Quinean reasons (Tymoczko, 1991), or even for Fregean reasons, then one would have no reason to talk of "intuition." For Quine, all objects are posited to obtain the most attractive overall theory.

2. Despite his conclusions, Quine is not a skeptic, because, as he says, he takes the original starting theory seriously.

3. The next two paragraphs are taken from Tymoczko, 1989. See also Tymoczko, 1990.

4. The first indication of such relativism is in Skolem's 1922 paper on axiomatized set theory. For a full reference to this and to other works by Skolem and for a discussion of the evolution of Skolem's views, see the fine article by Benacerraf (1985).

5. The connection between Putnam's argument about brains in a vat and the modified Skolem Paradox is a major focus of the paper mentioned in note 1 and is briefly discussed by Benacerraf (1985).

6. For a vigorous defense of Wittgenstein's position on the infinite, see Chap. 5 in Shanker, 1987.

REFERENCES

Anscombe, E., and P. Geach, eds.
1969 *Descartes, Philosophical Writings.* London: Nelson.

Benacerraf, P.
1985 "Skolem and the Skeptic." *The Aristotelian Society Proceedings* supp. vol. 59:85–116.

Browne, M. W.
1988 "Is a Math Proof a Proof If No One Can Check It?" *New York Times* Dec. 20: C1, C17.

Maddy, P.
1988 "Believing the Axioms." P. 1 and 2. *The Jornal of Symbolic Logic* 53:481–511, 736–64.

Putnam, H.
1981 *Reason, Truth and History.* Cambridge: Cambridge University Press.

Quine, W.
1960 *Word and Object.* Cambridge, Mass.: MIT Press.
1969 *Ontological Relativity and Other Essays.* New York: Columbia University Press.

Shanker, S.
1987 *Wittgenstein and the Turning Point in the Philosophy of Mathematics.* Albany, N. Y.: SUNY Press.

Thom, R.
1986 "'Modern' Mathematics: An Educational and Philosophic Error?" Pp. 67–78 in T. Tymoczko (ed.), *New Directions in the Philosophy of Mathematics.* Boston: Birkhauser.

Tymoczko, T.
1986 "The Four-Color Problem and its Philosophical Significance." Pp. 243–66 in T. Tymoczko (ed.), *New Directions in the Philosophy of Mathematics.* Boston: Birkhauser.
1989 "In Defense of Putnam's Brains." *Philosophical Studies* 57:281–97.
1990 "Brains Don't Lie." Pp. 195–213 in M. Roth and G. Ross (eds.), *Doubting: Contemporary Perspectives on Skepticism.* Dordrecht: Kluwer Academic Publishers.
1991 "Mathematics, Science and Ontology," *Synthese* 88:201–28.

Wittgenstein, L.
1956 *Remarks on the Foundations of Mathematics.* Oxford: Basil Blackwell.
1960 *The Blue and Brown Books.* Oxford: Basil Blackwell.
1976 *Wittgenstein's Lectures on the Foundations of Mathematics, 1939,* ed., C. Diamond, Sussex: Haverton Press.

5

Philosophical Problems of Mathematics
in the Light of Evolutionary Epistemology

INTRODUCTION

When one speaks of the foundations of mathematics or of its foundational problems, it's important to remember that mathematics is not an edifice which risks collapse unless it is seated on solid and eternal foundations that are supplied by some logical, philosophical, or extra-mathematical construction. Rather, mathematics ought to be viewed as an ever-expanding mansion floating in space, with new links constantly growing between previously separated compartments, while other chambers atrophy for lack of interested or interesting habitants. The foundations of mathematics also grow, change, and further interconnect with diverse branches of mathematics as well as with other fields of knowledge. Mathematics flourishes on open and thorny problems, and foundational problems are no exception. Such problems arose already in antiquity, but the rapid advance in the second half of the nineteenth century toward higher levels of abstraction and the recourse to the actual infinite by Dedekind[1] and Cantor all pressed for an intense concern with foundational questions. The discovery of irrational numbers, the use of negative numbers (from the Latin *negare*, literally, "to deny," or "to refuse"), the introduction of imaginary numbers, the invention of the infinitesimal calculus and the (incoherent) calculations with divergent series, and so forth, each of these novelties precipitated at their time uncertainties and resulted in methodological reflections. But starting with the creation of non-Euclidean geometries[2] and culminating in Cantor's theory of transfinite numbers, the *rate* at which new foundational problems presented themselves grew to the point of causing in some quarters a sense of crisis—hence the talk of a foundational crisis at the beginning of this century.

The philosophy of mathematics is basically concerned with systematic reflection about the nature of mathematics, its methodological problems, its relations to reality, and its applicability. Certain foundational inquiries, philosophical at the outset, were eventually internalized. Thus, the impetus resulting from philosophically motivated researches produced spectacular developments in the field of logic, with their ultimate absorption within mathematics proper. Today, the various descendants of foundational work, such as proof theory, axiomatic set theory, recursion theory, and so on, are part and parcel of the mainstream of mathematical research. This does not mean that the philosophy of mathematics has or ought to have withered away. On the contrary. Nowadays, many voices hail a renaissance in the philosophy of mathematics and acclaim its new vigor. Note also the current dynamic preoccupation by biologists and philosophers alike with foundational problems of biology. (A special journal, *Biology and Philosophy*, was created in 1986 to serve as a common forum.) By contrast, the mathematical community is rather insular, and most mathematicians now have a tendency to spurn philosophical reflections. Yet without philosophy we remain just stone heapers: "Tu peux certes raisonner sur l'arrangement des pierres du temple, tu ne toucheras point l'essentiel qui échappe aux pierres." ("You can certainly reason about the arrangement of the stones of the temple, but you'll never grasp its essence which lies beyond the stones," (my translation; Saint-Exupéry, 1948:256).

It is significant to notice that in the current literature on the philosophy of mathematics there is a marked shift towards an analysis of *mathematical practice* (cf. Feferman, 1985; Hersh, 1979; Kitcher, 1983; Kreisel, 1973; Resnik, 1975, 1981, 1982; Resnik and Kusher, 1987; Shapiro, 1983; Steiner, 1978a, 1987b, 1983; Van Bendegem, 1987). This is most refreshing, for it is high time that the philosophy of mathematics liberates itself from ever enacting the worn-out tetralogy of Platonism, logicism, intuitionism, and formalism. As Quine (1980:14) has pointed out, the traditional schools of the philosophy of mathematics have their roots in the medieval doctrines of realism, conceptualism and nominalism. Whereas the quarrel about universals and ontology *had* its meaning and significance within the context of medieval Christian culture, it is an intellectual scandal that some philosophers of mathematics can still discuss whether whole numbers exist or not. It *was* an interesting question to compare mathematical "objects" with physical objects as long as the latter concept was believed to be unambiguous. But, with the advent of quantum mechanics, the very concept of a physical object became more problematic than any mathematical concept.[3] In a nutshell, philosophy too has its paradigms, and a fertile philosophy of mathematics, like any other "philosophy of," must be solidly oriented towards the practice of its particular discipline and keep contact with actual currents in the philosophy of science. The purpose of this essay is to explore one such current in the philosophy of science, namely, *evolutionary epistemology*, with the tacit aim of hopefully obtaining some new insights concerning the nature of mathematical knowledge. This is not a reductionist program. But the search for new insights seem more fruitful than treading forever on the quicksand of neo-scholasticism and its offshoots. I concur with Wittgenstein that "a philosophical work consists essentially of elucidations" (1983:77).

THE MAIN TENETS OF EVOLUTIONARY EPISTEMOLOGY

Evolutionary Epistemology (EE) was independently conceived by Lorenz, a biologist; Campbell, a psychologist; and Vollmer, a physicist and philosopher. Though its origins can be traced to nineteenth-century evolutionary thinkers, EE received its initial formulation by Lorenz (1941) in a little-noticed paper on Kant. Christened in 1974 by Campbell and systematically developed in a book by Vollmer in 1975, evolutionary epistemology has quickly become a topic of numerous papers and books (see Campbell, Hayes, and Callebaut, 1987). In the opening paragraph of an essay in honor of Sir Karl Popper, where the term *evolutionary epistemology* appears for the first time, Campbell states:

> An evolutionary epistemology would be at minimum an epistemology taking cognizance of and compatible with man's status as a product of biological and social evolution. In the present essay it is also argued that evolution—even in its biological aspects—is a knowledge process, and that the natural-selection paradigm for such knowledge increments can be generalized to other epistemic activities, such as learning, thought, and science (Campbell, 1974:413).

My aim is to add mathematics to that list.

Riedl characterizes evolutionary epistemology as follows:

> In contrast to the various philosophical epistemologies, evolutionary epistemology attempts to investigate the mechanism of cognition from the point of view of its phylogeny. It is mainly distinguished from the traditional position in that it adopts a point of view outside the subject and examines different cognitive mechanisms comparatively. It is thus able to present objectively a series of problems [including the problems of traditional epistemologies, not soluble on the level of reason alone but soluble from the phylogenetic point of view]. (Reidel, 1984:220,1988:287.)

In an extensive survey article, Bradie (1986) introduced a distinction between two interrelated but distinct programs that go under the name of evolutionary epistemology. One one hand, there is an "attempt to account for the characteristics of cognitive mechanisms in animals and humans by a straightforward extension of the biological theory of evolution to those aspects or traits of animals which are the biological substrates of cognitive activity, e.g., their brains, sensory systems, motor systems, etc." (Bradie 1986:403). Bradie refers to this as the "Evolutionary Epistemology Mechanism program" (EEM). On the other hand, the EE Theory program, EET, "attempts to account for the evolution of ideas, scientific theories and culture in general by using models and metaphors drawn from evolutionary biology.[4] Both programs have their roots in 19th century biology and social philosophy, in the work of Darwin, Spencer and others" (Bradie, 1986:403). Popper is generally considered to be the main representative of the EET program, though Popper himself would not call himself an evolutionary epistemologist.[5] The great impetus to the EE Mechanisms program came from the work of Konrad Lorenz and his school of ethology. Through extensive studies of the behavior of animals in their natural habitat, Lorenz has deepened our understanding of the interplay between genetically determined and learned behavioral patterns. To Lorenz, the evolution of the cognitive apparatus is not different in kind from the evolution of organs. The same evolutionary mechanisms account for both. As Lorenz puts it in a famous passage:

> Just as the hoof of the horse, this central nervous apparatus stumbles over unforeseen changes in its task. But just as the hoof of the horse is adapted to the ground of the steppe which it

copes with, so our central nervous apparatus for organizing the image of the world is adapted to the real world with which man has to cope. Just like any organ, this apparatus has attained its expedient species-preserving form through this coping of real with the real during its genealogical evolution, lasting many eons (Lorenz, 1983:124).

In the fascinating 1941 paper already mentioned, Lorenz reinterpreted the Kantian categories of cognition in the light of evolutionary biology. By passing from Kant's *prescriptive epistemology* to an evolutionary *descriptive epistemology*, the category of *a priori cognition* is reinterpreted as the individual's inborn (a priori) *cognitive mechanisms* that have evolved on the basis of the species' a posteriori confrontation with the environment. In short, the phylogenetically a posteriori became the ontogenetically a priori. In the words of Lorenz, "The categories and modes of perception of man's cognitive apparatus are the natural products of phylogeny and thus adapted to the parameters of external reality in the same way, and for the same reason, as the horse's hooves are adapted to the prairie, or the fish's fins to the water" (1977:37, 1985:57).

Any epistemology worthy of its name must start from some postulate of realism: that there exists a real world with some organizational regularities. "In a chaotic world not only knowledge, but even organisms would be impossible, hence non-existent" (Vollmer, 1983:29). But the world includes also the reflecting individual. Whereas the idealist, to paraphrase Lorenz, looks only into the mirror and turns a back to reality, the realist looks only outwardly and is not aware of being a mirror of reality. Each ignores the fact that the mirror also has a nonreflecting side that is part and parcel of reality and consists of the physiological apparatus that has evolved in adaptation to the real world. This is the subject of Lorenz's remarkable book *Behind the Mirror*. Yet reality is not given to immediate and direct inspection. "Reality is veiled," to use the deft expression of d'Espagnat. But the veil can progressively be *transluminated*, so to speak, by conceptual modeling and experimentation. This is the credo of the working scientist. Evolutionary epistemology posits a minimal ontology, known under the name of *hypothetical realism*, following a term coined and defined by Campbell as follows:

> My general orientation I shall call hypothetical realism. An "external" world is hypothesized in general, and specific entities and processes are hypothesized in particular, and the observable

implications of these hypotheses (or hypostatizations, or reifica-
tions) are sought out for verification. No part of the hypotheses
has any "justification" or validity prior to, or other than through,
the testing of these implications. Both in specific and in general
they are always to some degree tentative (Campbell, 1959:156).

The reader is referred to the treatises by Vollmer (1987a, 1985;
1986) for a systematic discussion of evolutionary epistemology. See
also Ursua (1986) and Vollmer's (1984) survey article. Subsequently, I
will also draw on the insights furnished by the genetic (or developmental) epistemology of Piaget and his school (which I consider part
of the EE Mechanisms program), as well as on the work of Oeser
(1987, Oeser and Seitelberger 1988).

SOME PERENNIAL QUESTIONS IN THE
PHILOSOPHY OF MATHEMATICS

The Hungarian mathematician Alfréd Rényi has written a de-
lightful little book entitled *Dialogues on Mathematics*.[6] The first is a Socratic dialogue on the nature of mathematics, touching on some central themes in the philosophy of mathematics. From the following excerpts,[7] I shall extract the topics of our subsequent discussion.

Socrates: What things does a mathematician study? . . . [W]ould you
say that these things exist? . . . Then tell me, if there were no
mathematicians, would there be prime numbers, and if so,
where would they be?
Socrates: Having established that mathematicians are concerned with
things that do not exist in reality, but only in their thoughts, let
us examine the statement of Theaitetos, which you mentioned,
that mathematics gives us more trustworthy knowledge than
does any other branch of science.
Hippocrates: . . . [I]n reality you never find two things which are exactly the same; . . . but one may be sure that the two diagonals of
a rectangle are exactly equal . . . Heraclitus . . . said that everything which exists is constantly changing, but that sure knowledge is only possible about things that never change, for instance, the odd and the even, the straight line and the circle.
Socrates: . . . [W]e have much more certain knowledge about persons
who exist only in our imagination, for example, about characters
in a play, than about living persons . . . The situation is exactly

the same in mathematics.

Hippocrates: . . . But what is the use of knowledge of non-existing things such as that which mathematics offers?

Socrates: . . . How to explain that, as often happens, mathematicians living far apart from each other and having no contact, independently discover the same truth? I never heard of two poets writing the same poem . . . It seems that the object of [mathematicians'] study has some sort of existence which is independent of their person.

Socrates: But tell me, the mathematician who finds new truth, does he discover it or invent it?

Hippocrates: The main aim of the mathematician is to explore the secrets and riddles of the sea of thought. *These exist independently of the mathematician, though not from humanity as a whole* [italics mine].

Socrates: We have not yet answered the question: what is the use of exploring the wonderful sea of human thought?

Socrates: *If you want to be a mathematician, you must realize you will be working mostly for the future* [italics mine]. Now, let us return to the main question. We saw that knowledge about another world of thought, about things which do not exist in the usual sense of the word, can be used in everyday life to answer questions about the real world. Is this not surprising?

Hippocrates: More than that, it is incomprehensible. It is really a miracle.

Hippocrates: . . . [B]ut I do not see any similarity between the real world and the imaginary world of mathematics.

Hippocrates: . . . Do you want to say that the world of mathematics is a reflected image of the real world in the mirror of our thinking?

Socrates: . . . [D]o you think that someone who has never counted real objects can understand the abstract notion of number? . . . The child arrives at the notion of a sphere through experience with round objects like balls. Mankind developed all fundamental notions of mathematics in a similar way. These notions are crystallized from a knowledge of the real world, and thus it is not surprising but quite natural that they bear the marks of their origin, as children do of their parents. And exactly as children when they grow up become the supporters of their parents, so any branch of mathematics, if it is sufficiently developed, becomes a useful tool in exploring the real world.

Hippocrates: . . . Now we have found that the world of mathematics is nothing else but a reflection in our mind of the real world.

Socrates: . . . I tell you [that] the answer is not yet complete.
Socrates: *We have kept too close to the simile of the reflected image. A simile is like a bow—if you stretch it too far, it snaps* [italics mine].

Schematically, the key issues that emerge from the dialogue are the following:

1.   *Ontology*. In what sense can one say that mathematical "objects" exist?[8] If discovered, what does it mean to say that mathematical propositions are true independently of the knowing subject(s) and *prior* to their discovery?
2.   *Epistemology*. How do we come to know "mathematical truth" and why is mathematical knowledge considered to be certain and apodictic?
3.   *Applicability*. Why is mathematical knowledge applicable to reality?
4.   *Psychosociology*. If invented, how can different individuals invent the "same" proposition? What is the role of society and culture?

It has been stressed by Körner (1960) and by Shapiro (1983) that problem 3 is least adequately dealt with by *each* of the traditional philosophies of mathematics. As Shapiro rightly observes: "many of the reasons for engaging in philosophy at all make an account of the relationship between mathematics and culture a priority . . . Any world view which does not provide such an account is incomplete at best" (1983:524). To answer this challenge, there are voices that try to revive Mill's long-buried empiricist philosophy of mathematics, notwithstanding the obvious fact that mathematical propositions are not founded on sense impressions nor could any ever be refuted by empirical observations. How are we supposed to derive from experience that every continuous function on a closed interval is Riemann integrable? A more shaded empiricism has been advocated by Kalmar and Lakatos. Their position was sharply criticized by Goodstein (1970) and I fully agree with Goodstein's arguments (cf. the discussion in Lolli, 1982). In a different direction, Körner (1965) has sought an empiricist *justification* of mathematics via empirically verifiable propositions modulo translation of mathematical propositions into empirical ones. The problematics of translation apart, the knotty question of inductive justification "poppers" up again, and not much seems to be gained from this move. Though the road to empiricism is paved with good intentions, as with all such roads, the end point is the same. Yet the success of mathematics as a scientific tool is itself an empirical fact.

Moreover, empirical elements seem to be present in the more elementary parts of mathematics, and they are difficult to account for. To say that some mathematical concepts were formed by "abstraction" from experience only displaces the problem, for we still don't know how this process of abstraction is supposed to work. Besides, it is not the elementary part of mathematics that plays a fundamental role in the elaboration of scientific theories; rather, it is the totality of mathematics, with its most abstract concepts, that serves as a *pool* from which the scientist draws *conceptual schemes* for the elaboration of scientific theories. In order to account for this process, evolutionary epistemology starts from a minimal physical ontology, known as "hypothetical realism"; it just assumes the existence of an objective reality that is independent of our taking cognizance of it. Living beings, idealistic philosophers included, are of course part of objective reality. It is sufficient to assume that the world is nonchaotic; or, put positively, the world is assumed to possess organizational regularities. But I would not attribute to reality 'objective *relations*', 'quantitative *relations*', 'immutable *laws*', and so forth. All these are epistemic concepts and can only have a place within the frame of scientific theories. Some philosophers of mathematics have gone far beyond hypothetical realism and thereby skirt the pitfalls of both empiricism and Platonism, such as Ruzavin when he writes: "In complete conformity with the assertions of science, dialectical materialism considers mathematical objects as images, photographs, copies of the real quantitative relations and space forms of the world which surrounds us" (1977:193). But we are never told how, for instance, Urysohn's metrization theorem of topological spaces could reflect objective reality. Are we supposed to assume that through its pre-image in objective reality, Urysohn's theorem was already true before anybody ever thought of topological spaces? Such a position is nothing but Platonism demystified, and it would further imply that every mathematical problem is decidable independently of any underlying theoretical framework.

To reiterate: mathematics and objective reality are related, but the relationship is extremely complex, and no magic formula can replace patient epistemological analysis. We turn now to the task of indicating a direction for such an analysis from the point of view of evolutionary epistemology (cf. Vollmer, 1983).

MATHEMATICS AND REALITY

Many consider it a miracle—as Rényi had Hippocrates say—that mathematics is applicable to questions of the real world. In a famous article, Wigner expressed himself in a similar way: "the enormous usefulness of mathematics in the natural sciences is something bordering on the mysterious and . . . there is no rational explanation for it" (Wigner, 1960:2) It is hard to believe that our reasoning power was brought, by Darwin's process of natural selection, to the perfection which it seems to possess.

With due respect to the awe of the great physicist, *there is a rational explanation for the usefulness of mathematics* and it is the task of any epistemology to furnish one. Curiously, we'll find its *empirical basis* in the very evolutionary process which puzzled Wigner. Here is the theory that I propose.

The core element, the depth structure of mathematics, incorporates cognitive mechanisms, which have evolved like other biological mechanisms, by confrontation with reality and which have become genetically fixed in the course of evolution. I shall refer to this core structure as the *logico-operational component* of mathematics. Upon this scaffold grew and continues to grow the *thematic component* of mathematics, which consists of the specific content of mathematics. This second level is culturally determined and originated, most likely, from ritual needs. (The ritual origin of mathematics has been discussed and documented by numerous authors (cf. Seidenberg, 1981; Carruccio, 1977:10; Michaels, 1978, and their respective bibliographies). Notice that ritual needs were practical needs, seen in the context of the prevailing cultures; hence there is no more doubt about the practical origin of mathematics! Marchack (1972) has documented the presence of mathematical notations on bones dating to the Paleolithic of about thirty thousand years ago. This puts it twenty thousand years prior to the beginnings of agriculture; hence some mathematical knowledge was already available for the needs of land measurements, prediction of tides, and so forth. Given this remarkably long history, mathematics has been subjected to a lengthy cultural molding process akin to an environmental selection. Whereas the thematic component of mathematics is culturally transmitted and is in a continuous state of growth, the logico-operational component is based on genetically transmitted cognitive mechanisms and this is fixed. (This does not mean that the logico-operational level is ready for use at birth; it is still subject to an ontogenetic development.[9] The genetic program is an *open program* (Mayr, 1974:651–52) that is materialized in the phenotype under the influence of internal and external factors and is realized by stages in the development of the individual).

Let us look closer at the nature of cognitive mechanisms. Cognition is a fundamental physical process; in its simplest form it occurs at the molecular level when certain stereospecific configurations permit the aggregation of molecules into larger complexes. (There is no more anthropomorphism here in speaking of *molecular cognition* than in using the term *force* in physics.) As we move up the ladder of complexity, cognition plays a central role in prebiotic chemical evolution, and in the formation of self-replicating units. Here, in the evolution of macromolecules, "survival of the fittest" has a literal meaning: that *which fits, sticks* (chemically so!). That which doesn't fit, well, it just stays out of the game; it is "eliminated." These simple considerations should have a sobering effect when looking at more complicated evolutionary processes. The importance of cognition in the process of the self-organization of living matter cannot be overemphasized. Thus Maturana writes: "*Living systems are cognitive systems, and living as a process is a process of cognition*" (1980:13). What I wish to stress here is that there is a continuum of cognitive mechanisms, from molecular cognition to cognitive acts of organisms, and that some of these fittings have become genetically fixed and are transmitted from generation to generation. Cognition is not a passive act on the part of an organism but a dynamic process realized in and through *action*. Lorenz (1983:102) has perceptively pointed out that the German word for reality, *Wirklichkeit*, is derived from the verb *wirken*, "to act upon." The evolution of cognitive mechanisms is the story of successive fittings of the organism's actions upon the internal and external environments.

It is remarkable how complicated and well adapted inborn behavioral patterns can be, as numerous studies by ethologists have shown.

> Consider, for instance [writes Bonner], a solitary wasp. The female deposits her eggs in small cavities, adds some food, and seals off the chamber. Upon emergence the young wasp has never seen one of its own kind, yet it can walk, fly, eat, find a mate, mate, find prey, and perform a host of other complex behavioral patterns. This is all done without any learning from other individuals. It is awesome to realize that so many (and some of them complex) behavioral patterns can be determined by the genes. (1980:40)

Isn't this as remarkable as "that our reasoning power was brought, by Darwin's process of natural selection, to the perfection which it seems to possess?" (Recall the quote from Wigner.) From the rigid *single-choice behavior*, as in the case of the solitary wasp, through the evolution of *multiple-choice behavior*[10] and up to our capacity of *planned actions*, all intermediate stages occur and often concur.

> As behavior and sense organs became more complex [writes Simpson], perception of sensation from those organs obviously maintained a realistic relationship to the environment. To put it crudely but graphically, the monkey who did not have a realistic perception of the tree branch he jumped for was soon a dead monkey and therefore did not become one of our ancestors. Our perceptions do give true, even though not complete, representations of the outer world because that was and is a biological necessity, built into us by natural selection. If we were not so, we would not be here! We do now reach perceptions for which our ancestors had no need, for example, of X-rays or electric potentials, but we do so by translating them into modalities that are evolution-tested. (1963:84)

The nervous system is foremost a steering device for the internal and external coordination of activities. There is no such thing as an "illogical" biological coordination mechanism, else survival would not have been possible. "For survival," writes Oeser, "it is not the right images which count but the corresponding (re)actions" (1988:38). The coordinating activities of the nervous system proceed mostly on a subconscious level; we become aware of the hand that reaches out to catch a falling glass only at the end of the action. (It is estimated that from an input of 10^9 bits/sec, only 10^2 bits/sec reach consciousness.) Yet another crucial mechanism has evolved, known on the human level as *planned action*. It permits a choice of action or hypothetical reasoning: we can imagine, prior to acting, the possible outcome of an action and thereby minimize all risks. The survival value of anticipatory schemes is obvious. *When we form a representation for possible action, the nervous system apparently treats this representation as if it were a sensory input, hence processes it by the same logico-operational schemes as when dealing with an environmental situation* (cf. Shepard and Cooper, 1981, for some fascinating data). From a different perspective, Maturana and Varela express it this way: "all states of the nervous system are internal states, and the nervous system cannot make a distinction in its process of transformations between its internally and externally generated changes" (1980:131). Thus, the logical schemes in hypothetical representations are the same as the logical schemes in the coordination of actions, schemes that have been tested through eons of evolution and which by now are genetically fixed.

The preceding considerations have far reaching implications for mathematics. Under *"logico-mathematical schemes"* Piaget understands the cognitive schemes that concern groupings of physical objects, arranging them in order, comparing groupings, and so on. These basic premathematical schemes have a genetic envelope but mature by stages in the intellectual development of the individual. They are based on the equally genetically fixed *logico-operational schemes*, a term that I have introduced, as these schemes operate also on the nonhuman level. The logico-operational schemes form the basis of our logical thinking. As it is a fundamental property of the nervous system to function through recursive loops, any hypothetical representation that we form is dealt with by the same "logic" of coordination as in dealing with real life situations. Starting from the elementary logico-mathematical schemes, a hierarchy is established. Under the impetus of sociocultural factors, new mathematical concepts are progressively introduced, and each new layer fuses with the previous layers.[11] In *structuring new layers, the same cognitive mechanisms operate with respect to the previous layers as they operate with respect to an environmental input.* This may explain, perhaps, why the working mathematician is so prone to Platonistic illusions. The sense of reality that one experiences in dealing with mathematical concepts stems in part from the fact that in all our hypothetical reasonings, the object of our reasoning is treated in the nervous system by cognitive mechanisms, which have evolved through interactions with external reality (see also the quotation from Borel in note 16).

To summarize: mathematics does not reflect reality. But our cognitive mechanisms have received their imprimatur, so to speak, through dealing with the world. The empirical component in mathematics manifests itself not on the thematic level, which is culturally determined, but through the logico-operational and logico-mathematical schemes. As the patterns and structures that mathematics consists of are molded by the logico-operational neural mechanisms, these abstract patterns and structures acquire the status of *potential* cognitive schemes for forming abstract hypothetical world pictures. Mathematics is a singularly rich *cognition pool* of humankind from which schemes can be drawn for formulating *theories* that deal with phenomena that lie outside the range of daily experience and, hence, for which ordinary language is inadequate. Mathematics is structured by cognitive mechanisms, which have evolved in confrontation with experience, and, in its turn, mathematics is a tool for structuring domains of indirect experience. But mathematics is more than just a tool. *Mathematics is a collective work of art that derives its objectivity through social in-*

teraction. "A mathematician, like a painter or a poet, is a maker of patterns," wrote Hardy (1969:84). The metaphor of the weaver has been frequently evoked. But the mathematician is a weaver of a very special sort. When the weaver arrives at the loom, it is to find a fabric already spun by generations of previous weavers and whose beginnings lie beyond the horizons. Yet with the yarn of creative imagination, existing patterns are extended and sometimes modified. The weaver may only add a beautiful motif, or mend the web; at times, the weaver may care more about the possible use of the cloth. But the weaving hand, for whatever motive it may reach out for the shuttle, is the very *prehensile* organ that evolved as a grasping and branch clutching organ, and its coordinating actions have stood the test of an adaptive evolution.[12] In the mathematician, the artisan and artist are united into an inseparable whole, a unity that reflects the uniqueness of humankind as *Homo artifex.*

THE TRILEMMA OF A FINITARY LOGIC AND INFINITARY MATHEMATICS

In 1902, *L'enseignement mathématique* launched an inquiry into the working methods of mathematicians. The questionnaire is reproduced (in English translation) as appendix 1 in Hadamard (1945). Of particular interest is question thirty, which, among others, Hadamard addressed to Einstein. (No date for the correspondence is given, but I situate it in the forties when Hadamard was at Columbia University). Question thirty reads as follows:

It would be very helpful for the purpose of psychological investigation to know what internal or mental images, what kind of "internal world" mathematicians make use of, whether they are motor, auditory, visual, or mixed, depending on the subject which they are studying.

In his answer to Hadamard (1945, appendix 2:142–43), Einstein wrote:

(a) The words or the language, as they are written or spoken, do not seem to play any role in my mechanism of thought. The physical entities which seem to serve as elements in thought are certain signs and more or less clear images which can be "voluntarily" reproduced and combined . . .

(b) The above mentioned elements are, in my case, *of visual and some of muscular type.* Conventional words or other signs have to be sought for laboriously only in a secondary stage, when the mentioned associative play is sufficiently established and can be reproduced at will [italics mine].

In the previous section I have discussed the core structures of mathematics which consist of the logico-operational schemes for the coordination of actions. Throughout the evolution of hominoids, the coordinating mechanisms of the hand and eye played a particularly important role, leading to the feasibility of extensive use of tools and, thereby, to further cortical developments. It is therefore not surprising that in dealing with concepts, where the same neural mechanisms are involved, visual and traces of kinesthetic elements manifest themselves in consciousness, as Einstein's testimonial confirms.

The world of our immediate actions is *finite,* and the neural mechanisms for anticipatory representations were forged through dealing with the finite. Formal logic is not the source of our reasoning but only *codifies* parts of the reasoning processes. But whence comes the feeling of safety and confidence in the soundness of the schemes that formal logic incorporates? To an evolutionary epistemologist, logic is not based on conventions; rather, we look for the biological substrata of the fundamental schemes of inference. Consider for instance *modus ponens:*

$$A \to B$$

$$A$$

$$\overline{}$$

$$\therefore B.$$

If a sheep perceives only the muzzle of a wolf, it promptly flees for its life. Here, "muzzle → wolf" is "wired" into its nervous system. Hence the mere sight of a muzzle—the muzzle of any wolf, not just the muzzle of a particular wolf—results in "inferring" the presence of a wolf. Needless to say, such inborn behavioral patterns are vital. For related examples, see Lorenz (1973) and Riedl (1979). The necessary character of logic, qua codified logico-operational schemes, thus receives a coherent explanation in view of its phylogenetic origin. It follows furthermore that *as far as logic is concerned, finitism does not need any further philosophical justification.* It is biologically imposed. The situation is different with respect to the *thematic component* of mathematics. Once the cultural step was taken in inventing number words and symbols which can be indefinitely extended, mathematics proper, as the science of the infinite, came into being. The story of the early philosophical groping with mathematical and possible physical infinity is well known.[13] When at last full citizenship was conferred on the *actual infinite*—de facto by Kummer and Dedekind; *de jure* by Cantor and Zermelo—an intense preoccupation with foundational problems was set in motion.[14] The first school to emerge was logicism à la Frege and Russell. "The logicistic thesis is," writes Church, "that logic and mathematics are related, not as two different subjects, but as earlier and later parts of the same subject, and indeed in such a way that mathematics can be obtained from pure logic without the introduction of additional primitives or additional assumptions" (Church, 1962:186). *Had the logicist programme succeeded, then infinitary mathematics, a cultural product, would have received a finitary foundation in finitary, biologically based logic.* But as early as 1902, Keyser already showed that mathematical induction required an axiom of infinity, and, finally, Russell had to concede that such an axiom (plus the axiom of reducibility) had to be added to his system. Thus, *the actual infinite is the rock upon which logicism foundered.* Still, the efforts of the logicist school were not in vain, as Church has pointed out: "it does not follow that logicism is barren of fruit. Two important things remain. One of these is the reduction of mathematical vocabulary to a surprisingly brief list of primitives, all belonging to the vocabulary of pure logic. The other is the basing of all existing mathematics on one comparatively simple unified system of axioms and rules of inference" (1962:186).[15]

The second attempt in finitist foundations for mathematics was undertaken by Hilbert in his famous program. It may not be inopportune to stress that Hilbert never maintained seriously that mathematics is devoid of content, and his oft-cited *mot d'esprit* that "mathematics is a game played according to certain simple rules with meaningless marks on paper" has regrettably resulted in unwarranted philosophical extrapolations. Hilbert's formalist program is a *technique*, a device, for proving the consistency of infinitary mathematics by finitistic means. In the very article in which he outlines his program, Hilbert said the following concerning Cantor's theory of transfinite numbers: "This appears to me the most admirable flower of the mathematical intellect and in general one of the highest achievements of purely rational human activity" (1967:373). A meaningless game? Hardly!

Through formalization of thematic mathematics, Hilbert proposed that "contentual inference [be] replaced by manipulation of signs according to rules" (1967:381). This *manipulation* (*manus*, literally "hand"), this *handling* of inscriptions in the manner one handles physical objects would be founded, from the perspective of evolutionary epistemology, on the safe *logico-operational schemes for dealing with the finite*. It was a magnificent program, and though in view of Gödel's incompleteness theorem it could not be carried out as originally conceived, its offshoot, proof theory, is a major flourishing branch of mathematical logic. Thus, the contributions of logicism and Hilbert's program are of lasting value. As to the original intent, we just have to accept that *one cannot catch an infinite fish with a finite net!* Thus there remain three alternatives:

1.   Use an infinite net, say of size $\in^0$ (Gentzen)
2.   Eat only synthetic fish (Brouwer)
3.   Be undernourished and settle for small fish (strict finitism).

Chacun a son goût!

INVENTION VERSUS DISCOVERY

"But tell me," asked Socrates in Rényi's dialogue, "the mathematician who finds new truth, does he discover it or invent it?" We all know that a time-honored way to animate an after-dinner philosophical discussion is to ask such a question. People agree that following common usage of language, Columbus did not invent America, nor did Beethoven discover the ninth symphony. But when a new drug has been synthesized we commonly speak of a discovery, though the molecule never existed anywhere prior to the creative act of its synthesizers. Hadamard, in the introduction to his book *The Psychology of Invention in the Mathematical Field*, observes that "there are plenty of examples of scientific results which are as much discoveries as inventions," and thus he prefers not to insist on the distinction between invention and discovery (1945:xi). Yet there are philosophers of mathematics who are committed to an essential distinction between discovery and invention. To the intuitionist, mathematical propositions are mental constructions and, as such, could not result from a discovery. The Platonist, on the other hand, believes "that mathematical reality lies outside us, that our function is to discover or *observe* it," as Hardy (1969:123) put it. The conventionalist, though for different reasons, would side with the intuitionist and consider mathematics to be invented. Apparently, neither logicism nor formalism is committed to a discovery/invention dichotomy. Is the debate about invention versus discovery an idle issue, or can one use the commonsense distinction between the two terms in order to elucidate the distinct components in the growth of mathematical knowledge? Let us examine the issue through a standard example. I propose to argue that: (1) the *concept* of 'prime number' is an invention; (2) the *theorem* that there are infinitely many prime numbers is a discovery. (N.b., Euclid's formulation, book 9, prop. 20, reads: "Prime numbers are more than any assigned multitude of prime numbers.")

Why should the concept of prime number be considered an invention, a purely creative step that need not have been taken, while contrariwise it appears that an examination of the factorization properties of the natural numbers leads immediately to the "discovery" that some numbers are composite and others are not, and this looks like a simple "matter of fact"? Weren't the prime numbers already there, tucked away in the sequence of natural numbers prior to anyone noticing them? Now things are not that simple. First of all, the counting numbers, like other classificatory schemes, did not make a sudden appearance as an indefinitely extendable sequence. Some cultures never went beyond coining words for the first few whole numbers. There are even languages destitute of pure numeral words. But even in cultures with a highly developed arithmetic, like the ones in ancient Babylonia, Egypt, or China, the concept of prime number was absent. Mo (1982) has shown how mathematicians in ancient China, though lacking the concept of prime number, solved problems such as reduction of fractions to lowest terms, addition of fractions, and finding Pythagorean triplets. Could it reasonably be said that the Chinese just missed "discovering" the prime numbers, and that so did the Babylonians and the Egyptians, in spite of their highly developed mathematical culture extending over thousands of years? I don't think so. In retrospect it seems to us that there was some sort of necessity that the concept of prime number be stumbled upon. But this is a misleading impression. *Evolution, be it biological or cultural, is opportunistic.* Much of our modern mathematics would still stay intact if the concept of prime number were lacking, though number theory and, hence, portions of abstract algebra would be different. There are $2^{\aleph_0}$ subsets of $\mathbb{N}$ of which only $\aleph_0$ can be defined by any linguistic means. We neither discover nor invent any one of these subsets separately. But when the inventive step was taken in formulating the *concept* of prime numbers, one of the subsets of $\mathbb{N}$ was singled out (that is, to serve as a *model,* in modern terminology). Some historians of mathematics attribute to the Pythagoreans certain theorems involving primes, but it is more likely that the concept of prime number is of a later date. It is conceivable that there is a connection between cosmological reflections about the ultimate constituents of matter by the Greek atomists and thoughts about numerical atoms, that is, prime numbers. Whatever the tie may be, one thing is certain: the invention of mathematical concepts is tied to culture. As White (1956) affirmed contra Platonistic doctrines, the "locus of mathematical reality is cultural tradition." The evolution of mathematical concepts can be understood only in the appropriate sociocultural context.[16]

Let us note that concepts can be defined explicitly, as in the case of prime numbers, or implicitly, by a system of axioms, like the concept of a group. In either case it is an inventive act. Theorems, on the other hand, have more the character of a discovery, in the sense that one *discovers* a road linking different localities. Once certain concepts have been introduced and, so to speak, are already there, it is a matter of discovering their connection, and this is the function of proofs. To come back to the theorem that no finite set of primes can contain all the prime numbers, it has the character of a discovery when one establishes a *road map* (Goodstein, 1970) linking "set of primes," "number of elements," and so on, to yield a path to the conclusion. A proposed path may or may not be valid, beautiful, or interesting. But to say that a proof renders a proposition "true" is as metaphorical as when one claims to have found a "true" path. *It seems best to dispense altogether with the notion of mathematical truth.* (This has no bearing on the *technical* metamathematical notion of 'truth' in the sense of Tarski.) Gone is, then, too the outdated Aristotelian conception of 'true axioms'. (Think of Euclidean and non-Euclidean geometries.) Such a "no truth" view also resolves the infinite regress involved in the apparent flow of truth from axioms to theorems that Lakatos (1962) endeavored to eliminate by an untenable return to empiricism. The creative work of the mathematician consists of inventing concepts and developing methods permitting one to chart paths between concepts.[17] This is how mathematics grows in response to internal and external *problems* and results in an edifice that is beautiful and useful at the same time.

RECAPITULATION AND CONCLUDING REMARKS

The evolutionary point of view dominated this essay, both in its metaphorical as well as in its strict biological sense. I started with the view of mathematics as an evolving mansion, foundations included. In harmony with the current emphasis in the philosophy of mathematics on actual mathematical practice, one of my chief concerns was to elucidate the relationship between mathematics and external reality. Though I rejected empiricism as an inadequate philosophy of mathematics, I endeavored to account for the empirical components in mathematics whose presence is clearly felt but which are difficult to locate. Mathematics is a science of structures, of abstract patterns (cf. Resnik, 1981, 1982). It is a human creation, hence it is natural to look for biological as well as sociocultural factors that govern the genesis of mathematical knowledge. The success of mathematics as a cognitive tool leaves no doubt that some basic biological mechanisms are involved. The acquisition of knowledge by organisms, even in its simplest form, presupposes mechanisms that could have evolved only under environmental pressure. Evolutionary epistemology starts from the empirical fact that our cognitive apparatus is the result of evolution and holds that our world picture must be appropriate for dealing with the world, because otherwise survival would not have been possible (cf. Bollmer, 1975:102). Indeed, it is from the coordination of *actions* in dealing with the world that anticipatory schemes of action have evolved; these, in turn, are at the root of our logical thinking. Thus, the *phylogenetically* but not individually empirical element manifests itself in our logico-operational schemes of actions, which lie at the root of the elementary logico-mathematical operations as studied by Piaget. On the other hand, the *content* of mathematical theories is culturally determined, but the overall mathematical formation sits on the logico-operational scaffold. Mathematics is thus seen as a two-tiered web: a *logico-operational level* based on cognitive mechanisms which have become fixed in adaptation to the world, and a *thematic level* determined by culture and social needs and hence in a continuous process of growth. This special double-tiered structure endows mathematics in addition to its artistic value with the function of a *cognition pool* which is singularly suitable beyond ordinary language for formulating scientific concepts and theories.

In the course of my discussion I also reassessed the rationale of logicism and Hilbert's program. Of the traditional philosophies of mathematics, only Platonism is completely incompatible with evolutionary epistemology. "How is it that the Platonistic conception of mathematical objects can be so convincing, so fruitful and yet so clearly false?" writes Paul Ernest in a review (1983).[18] I disagree with Ernest on only one point: I do not think that Platonism is fruitful. As a matter of fact, Platonism has *negative effects* on research by blocking a dynamical and dialectic outlook. Just think of set theorists who keep looking for "the true axioms" of set theory, and the working mathematician who will not explore on equal footing the consequences of the negation of the continuum hypothesis as well as the consequences of the affirmation of the continuum hypothesis. For the same reason too many logicians still ignore paraconsistent and other "deviant" (!) logics. Like the biological theory of preformation—which is just another side of the same coin—Platonism has deep sociological and ideological roots. What Dobzhansky had to say about the preformist way of thinking applies *mutatis mutandis* to Platonism:

> The idea that things are preformed, predestined, just waiting around the corner for their turn to appear, is pleasing and comforting to many people. Everything is destiny, fate. But to other people predestination is denial of freedom and novelty. They prefer to think that the flow of events in the world may be changed creatively, and that new things do arise. The influence of these two types of thinking is very clear in the development of biological theories. (1955:223)

And so it is in the philosophy of mathematics.

Starting with the misleading metaphor of mathematical truth, Platonists graft onto it the further misleading metaphor of mathematical objects as physical objects to which "truth" is supposed to apply. Metaphors are illuminating, but when metaphors are stacked one upon the other without end, the result is obscurity and finally obscurantism. I frankly confess that I am absolutely incapable of understanding what is meant by "ontological commitment" and the issue of the "existence of abstract objects," and I begin to suspect that the emperor wears no clothes. No, there are no preordained, predetermined mathematical "truths" that lie just out or up there. Evolutionary thinking teaches us otherwise.

> *Caminante, son tus huellas*
> *el camino y nada mas;*
> *caminante, no hay camino,*
> *se hace camino al andar.*
>   (Antonio Machado)

> Walker, just your footsteps
> are the path and nothing more;
> walker, no path was there before,
> the path is made by act of walking.

NOTES

This is an expanded version of talks presented at the International Congress "Communication and Cognition. Applied Epistemology," Ghent, December 6–10, 1987; at the Logic Seminar of the Kurt-Gödel-Gesellschaft, Technical University, Vienna, May 30, 1988; and at the Séminaire de Philosophie et Mathématique, Ecole Normale Supérieure, Paris, November 7, 1988. I thank the various organizers and participants for numerous stimulating conversations and discussions.

1. Concerning the role of Dedekind, often neglected in foundational discussions, see Edwards (1983). In his review of Edwards's paper, Dieudonné makes the following significant observation: "Dedekind broke entirely new ground in his free use of 'completed' infinite sets as single objects on which one could compute as with numbers, long before Cantor began his work on set theory" (Mathematical Reviews 84d:01028).

2. "With non-Euclidean geometry came into being a new state of mind which impressed its spirit of freedom on the whole development of modern mathematics" (Toth, 1986:90; this fascinating essay deals in considerable depth with the epistemological problem of non-Euclidean geometries).

3. From a current point of view, physical objects are considered as events or states that rest unaltered for a nonnegligible time interval. Though "event" and "state" *refer* to reality, in order to *speak* of them one needs the mathematical apparatus incorporated in physical theories. Thus one ends up again with mathematical concepts. Hence it is futile to look at mathematical concepts as *objects* in the manner of physical objects and then, to crown it all, relegate them to a Platonic abode. For a further discussion of ontological questions concerning physical objects, see Dalla Chiara, 1985; Dalla Chiara and Toraldo di Francia, 1982; Quine, 1976.

4. Vollmer (1987a, 1987b) also stresses the difference between EE à la Lorenz as a *biological theory of the evolution of cognitive systems* and EE à la Popper as a theory of the *evolution of scientific ideas*.

5. Cf. Popper's disclaimer in Riedl et al., 1987:24. In Popper's philosophy, factual knowledge cannot serve as a basis for an epistemology, whereas evolutionary epistemology is committed to an "irresolvable nexus between empirical knowledge and metatheoretical reflections," following Vollmer. Moreover, the great strides of science in the last fifty years are due to ever-refined experimental techniques and technologies coupled with piecemeal modeling, rather than to the elaboration of grand theories. When one peeks into a modern reasearch institute, one scarcely finds scientists engaging in a grandiose search for bold hypotheses and a frantic pursuit of refutations, but rather humbly approaching "nature with the view, indeed, of receiving information from it, not, however, in the character of a pupil, who listens to all that his master chooses to tell him, but in that of a judge, who compels the

witness to reply to those questions which he himself thinks fit to propose" (Kant, 1934:10–11)

6. See Rényi, 1967. The booklet contains three dialogues: (1) "A Socratic Dialogue on Mathematics," whose protagonists are Socrates and Hippocrates; (2) "A Dialogue on the Applications of Mathematics," featuring Archimedes and Hieron; and (3) "A Dialogue on the Language of the Book of Nature," whose chief character is Galileo.

7. I have excerpted this material from pages 7–25 in A. Rényi, *Dialogues on Mathematics* (San Francisco: Holden Day, 1967). These excerpts are quoted with the kind permission of the publishers.

8. Aristotle, 1941 discussed the difficulties with the Platonist notion of mathematical *objects* and their existence. See *Metaphysics*, book 13, chap. 1–3, 1076a 33–1078b 6.

9. We owe much of our understanding of the ontogenetic development of the various logico-mathematical schemes to the work of Piaget and his school. Compare Müller, 1987:102–6, for a succinct summary of Piaget's theory. Note that much though not all of Piaget's (onto)genetic epistemology is compatible with evolutionary epistemology. See the discussions by Apostel (1987) and Oeser (1988:40, 165).

10. These terms are given by Bonner (1980). Of particular importance is Bonner's extension of the concept of 'culture', which he defines as follows:

> By culture I mean the transfer of information by behavorial means, most particularly by the process of teaching and learning. It is used in a sense that contrasts with the transmission of genetic information passed by direct inheritance of genes. The information passed in a cultural fashion accumulates in the form of knowledge and tradition, but the stress of this definition is on the mode of transmission of information, rather than its result. *In this simple definition I have taken care not to limit it to man.* (1980:10; italics mine)

11. It is a "fundamental principle of neuro-epistemology," writes Oeser, "that each new cognitive function results from an integration with previously formed and already existing functions" (1988:158).

12. The evolution of the hand as a prehensile organ not only enabled humans to grasp physical objects but led concomitantly to neural mechanisms enabling them to grasp relationships between objects. This is the path from prehension to comprehension, or in German, as Lorenz has pointed out, from *greifen* (to grasp), via *begreifen* (to understand), to *Begriff* (concept) (see Lorenz, 1973:192–94; Vollmer, 1975:104–5; Oeser and Seitelberger, 1988:159). From a neurophysiological point of view, notice the large area of the cortical maps of the hands (see Granit, 1977:64–65).

13. For a collection of most of the relevant passages in Aristotle, see Apostel (1952). Augustine had no qualms about the *actual infinite* in mathematics, to wit: "Every number is defined by its own unique character, so that no number is equal to any other. They are all unequal to one another and different, and the individual numbers are finite but *as a class they are infinite*" (1984:496; my italics).

14. The "paradoxes" played only a minor role in this process and none in the case of Frege. For a discussion, compare Garciadiego, 1986 and the review by Corcoran in *Mathematical Reviews* 1988 (88a:01026).

15. However, there is no *unique* set theory with a unique underlying logic from which all presently known mathematics can be derived. (Just recall the numerous independence results and the needs of category theory.) Moreover, when one examines actual mathematical practice, the deficiencies of "standard logic" are apparent, as Corcoran (1973) has perspicaciously pointed out. Furthermore, cognitive psychologists and workers in artificial intelligence are keenly aware of the fact that our current schemes of formal logic are inapplicable for analyzing actual reasoning processes. (cf. Gardner, 1985:368–70 and the references cited therein). Much work needs to be done in developing a logic of actual reasoning.

16. For a further discussion, see White, 1956; Wilder, 1981. And Borel adds the following perceptive observation:

> [W]e tend to posit existence on all those things which belong to civilization or culture in that we share them with other people and can exchange thoughts about them. Something becomes objective (as opposed to "subjective") as soon as we are convinced that it exists in the minds of others in the same form as it does in ours, and that we can think about it and discuss together. Because the language of mathematics is so precise, it is ideally suited to defining concepts for which such a consensus exists. In my opinion, that is sufficient to provide us with a *feeling* of an objective existence, of a reality of mathematics (1983:13)

17. A radioscopy of mathematical proofs reveals their logical structure, and this aspect has traditionally been overemphasized at the expense of seeing the meat and flesh of proofs. The path between concepts not only has a logical part which serves to *convince*; such paths also establish interconnections which *modify* and *illuminate* complexes of mathematical ideas, and this is how proofs differ from derivations.

18. Similarly, Machover writes concerning Platonism: "The most remarkable thing about this utterly incredible philosophy is its success" (1983:4). And further down: "The clearest condemnation of Platonism is not so much its belief in the occult but its total inability to account for constructive mathematics" (1983:5).

REFERENCES

Apostel, L.
1987 "Evolutionary Epistemology, Genetic Epistemology, History and Neurology." Pp. 311–26 in W. Callebaut and R. Pinxten (1987).

Apostle, H. G.
1952 *Aristotle's Philosophy of Mathematics.* Chicago: University of Chicago Press.

Aristotle
1941 *The Basic Works of Aristotle.* New York: Random House.

Augustine
1984 *Concerning the City of God against the Pagans,* a new translation by H. Bettenson, Harmondsworth: Penguin.

Beth, E. J., and J. Piaget
1966 *Mathematical Epistemology and Psychology.* Dordrecht: D. Reidel.

Bonner, J. T.
1980 *The Evolution of Culture in Animals.* Princeton: Princeton University Press.

Borel, A.
1983 "Mathematics, Art and Science." *The Mathematical Intelligencer* 5, 4:9–17.

Bradie, M.
1986 "Assessing Evolutionary Epistemology." *Biology and Philosophy* 1:401–59.

Callebaut, W., and R. Pinxten, eds.
1987 *Evolutionary Epistemology: A Multiparadigm Program.* Dordrecht: D. Reidel.

Campbell, D. T.
1959 "Methodological Suggestions from a Comparative Psychology of Knowledge Processes." *Inquiry* 2:152–82.
1974 "Evolutionary epistemology." Pp. 413–63 in P. A. Schilpp (ed.), *The Philosophy of K. Popper,* pt. 1. La Salle: Open Court.

Campbell, D. T., C. Heyes, and W. Callebaut
1987 "Evolutionary Epistemology Bibliography." Pp. 402–31 in Callebaut and R. Pinxten (1987).

Carruccio, E.
1977 *Appunti di storia delle matematiche, della logica, della metamatematica.* Bologna: Pitagora Editrice.

Church, A.
1962 "Mathematics and Logic." Pp. 181–86 in E. Nagel, P. Suppes and A. Tarski, (eds.), *Logic, Methodology and Philosophy of Science.* Proceedings, 1960 International Congress on Logic, Methodology and Philosophy of Science, Stanford. Stanford: Stanford University Press.

Corcoran, J.
1973 "Gaps between Logical Theory and Mathematical Practice." Pp. 23–50 in M. Bunge (ed.), *The Methodological Unity of Science*. Dordrecht: D. Reidel.

d'Espagnat, B.
1979 *A la recherche du réel: Le regard d'un physicien*. Paris: Gauthier-Villard

d'Espagnat, B. and M. Paty
1980 "La physique et le réel". *Bulletin de la société française de philosophie* 84:1–42.

Dalla Chiara, M. L.
1985 "Some Foundational Problems in Mathematics Suggested by Physics." *Synthese* 62, 2:303–15.

Dalla Chiara, M. L. and G. T. di Francia
1982 "Consideraciones ontologicas sobre los objectos de la fisica moderna." *Analisis Filosofico* 2:35–46.

Dobzhansky, H. M.
1955 *Evolution, Genetics, and Man*. New York: John Wiley.

Edwards, H. M.
1983 "Dedekind's Invention of Ideals." *Bulletin of the London Mathematical Society*. 15:8–17.

Ernest, P.
1983 *Mathematical Reviews* 83K(00010):4381

Feferman, S.
1985 "Working Foundations." *Synthese* 62, 2:229–54.

Garciadiego, A. R.
1986 "On Rewriting the History of the Foundations of Mathematics." *Historia Mathematica* 13:38–41.

Gardner, H.
1985 *The Mind's New Science: A History of the Cognitive Revolution*. New York: Basic Books.

Goodstein, R. L.
1970 "Empiricism in Mathematics." *Dialectica* 23:50–57.

Granit, R.
1977 *The Purposive Brain*. Cambridge, Mass.: MIT Press.

Hadamard, Jacques S.
1945 *The Psychology of Invention in the Mathematical Field*. Princeton: Princeton University Press.

Hardy, G. H.
1940 *A Mathematician's Apology*. Cambridge: Cambridge University Press.

106   *Philosophical Perspectives*

Hersh, R.
1979 "Some proposals for reviving the philosophy of mathematics." *Advances in Mathematics* 31:31–50.

Hilbert, D.
1967 "On the Infinite." Pp. 369–92 in J. Van Heijenoort (ed.), *From Frege to Gödel.* Cambridge, Mass.: Harvard University Press.

Kant, E.
1934 *Critique of Pure Reason* second ed., translated by J. H. D. Meiklejohn. London: Everyman's Library.

Kaspar, R.
1983 "Die biologischen Grundlagen der evolutionären Erkenntnistheorie." Pp. 125–45 in K. Lorenz et al. (1983).

Keyser, C. J.
1902 "Concerning the Axiom of Infinity and Mathematical Induction." *Bulletin of the American Mathematical Society.* 9:424–34.

Kitcher, P.
1981 "Mathematical Rigor—Who Needs It?" *Noûs* 15:469–93.
1983 *The Nature of Mathematical Knowledge.* New York: Oxford University Press.

Körner, S.
1960 *The Philosophy of Mathematics.* London: Hutchins.
1956 "An Empiricist Justification of Mathematics." Pp. 222–27 in Y. Bar-Hillel (ed.), *Logic, Methodology and Philosophy of Science.* Proceedings, 1964, International Congress of Logic, Methodology and Philosophy of Science, Jerusalem. Amsterdam: North-Holland.

Kreisel, G.
1973 "Perspectives in the Philosophy of Mathematics." Pp. 255–77 in P. Suppes (ed.), *Logic, Methodology and Philosophy of Science IV.* Amsterdam, North-Holland.

Lakatos, I.
1962 "Infinite Regress and the Foundations of Mathematics." *Aristotelian Society Proceedings* 36:155–84.

Lolli, G.
1982 "La dimostrazione in matematica: Analisi di un dibattido." *Bolletino Unione Matematica Italiana* Vol. 61, 1:197–216.

Lorenz, K.
1983 "Kants Lehre vom Apriorischen in Lichte gegenwärtiger Biologie." Pp. 95–124 in Lorenz and Wuketits, (1983).

Lorenz, K. and F. M. Wuketits, eds.
1983 *Die Evolution des Denkens.* Munich: R. Pieper.

Machover, M.
1983 "Towards a new philosophy of mathematics." *British Journal for Philosophy of Science* 34:1–11.

Marshack, A.
1972 *The Roots of Civilization*. London: Weiderfeld & Nicolson.

Maturana, H. R.
1980 "Biology of Cognition." Pp. 5–57 in Maturana and Varela (1980).

Maturana, H. R. and F. J. Varela
1980 *Autopoiesis and Cognition*. Dordrecht: D. Reidel.

Mayr, E.
1974 "Behavioral Programs and Evolutionary Strategies." *American Scientist* 62:650–659.

Michaels, A.
1978 *Beweisverfahren in der vedischen Sakralgeometrie*. Wiesbaden: Franz Steiner.

Mo, S. K.
1982 "What Do We Do If We Do Not Have the Concept of Prime Numbers? A Note on the History of Mathematics in China." *Journal of Mathematical Research Expositions* 2:183–87. In Chinese; English summary in *Mathematical Reviews* 84i:01024.

Müller, H. M.
1987 *Evolution, Kognition und Sprache*. Berlin: Paul Parey.

Oeser, E.
1987 *Psychozoikum: Evolution und Mechanismus der menschlichen Erkenntnisfähigkeit*. Berlin: Paul Parey.

Oeser, E., and F. Seitelberger
1988 *Gehirn, Bewusstsein und Erkenntnis*. Berlin: Paul Parey.

Piaget, J.
1967 *Biologie et Connaissance*. Paris: Gallimard.
1970 *Genetic Epistemology*. New York: Columbia University Press.
1971 *Biology and Knowledge: An Essay on the Relations between Organic Regulations and Cognitive Processes*. Chicago: University of Chicago Press. (English translation by Beatrix Walsh of Piaget, 1967).

Plotkin, H. C., ed.
1982 *Learning, Development and Culture: Essays in Evolutionary Epistemology*. New York: John Wiley.

Popper, K. R.
1972 *Objective Knowledge: An Evolutionary Approach*. Oxford: Oxford University Press.

Quine, W. V. O.
1976   "Whither Physical Objects?." Pp. 497–504 in R. S. Cohen, P. K. Feyer-
       abend and M. W. Wartofsky (eds.), *Essays in Memory of Imre Lakatos*.
       Dordrecht: D. Reidel.
1980   "On What There Is." Pp. 1–19 in W. V. O. Quine, *From a Logical Point
       of View*. Cambridge, Mass.: Harvard University Press.

Rényi, A.
1967   *Dialogues on Mathematics*. San Francisco: Holden-Day.

Resnik, M. D.
1975   "Mathematical Knowledge and Pattern Cognition." *Canadian Journal
       for Philosophy* 5:25–39.
1981   "Mathematics as a Science of Patterns: Ontology and Reference."
       Noûs 15:529–50.
1982   "Mathematics as a Science of Patterns: Epistemology." *Noûs*
       16:95–105.

Resnik, M. D., and D. Kushner
1987   "Explanation, Independence and Realism in Mathematics." *British
       Journal for Philosophy of Science* 38:141–58.

Riedl, R.
1979   *Biologie der Erkenntnis: Die stammgeschichtlichen Grundlagen der Ver-
       nunft*. Berlin: Paul Parey.
1984   *Biology of Knowledge: The Evolutionary Basis of Reason*. Chichester:
       John Wiley. (English translation by Paul Feulkes of Riedl, 1979.)

Riedl, R. and F. M. Wuketits, eds.
1987   *Die evolutionäre Erkenntnistheorie*. Berlin: Paul Parey.

Ruzavin, G. I.
1977   *Die Natur der mathematischen Erkenntnis*. Berlin: Akademie Verlag.

Saint-Exupéry, A.
1948   *Citadel*. Paris: Gallimard.

Seidenberg, A.
1962   "The ritual origin of counting." *Archives of the History of the Exact Sci-
       ences* 2:1–40.
1977   "The origin of mathematics." *Archives of the History of the Exact Sci-
       ences* 18:301–42.
1981   "The ritual origin of the circle and square." *Archives of the History of
       the Exact Sciences* 25:269–327.

Shapiro, S.
1983   "Mathematics and Reality." *Philosophy of Science* 50:523–548.

Shepard, R. N., and L. A. Cooper
1981   *Mental Images and Their Transformations*. Cambridge, Mass.: MIT
       Press.

Simpson, G. G.
1963 "Biology and the Nature of Science." *Science* 139:81–88.

Steiner, M.
1978a "Mathematics, Explanation, and Scientific Knowledge." *Noûs* 12:17–28.
1978b "Mathematical Explanation." *Philosophical Studies,* 34:135–51.
1983 "Mathematical realism." *Noûs* 17:363–85.

Toraldo di Francia, G.
1978 "What is a physical object?" *Scientia* 113:57–65.

Toth, I.
1986 "Mathematische Philosophie und hegelsche Dialektik." Pp. 89–182 in M. J. Petry, (ed.), *Hegel und die Naturwissenschaften.* Stuttgart: Frommann-Holzboog.

Ursua, N.
1986 "Conocimento y realidad: Aproximacion a una hipotesis." *Theoria* 2, 5–6;461–502.

Van Bendegem, J. P.
1987 "Fermat's Last Theorem Seen as an Exercise in Evolutionary Epistemology." Pp. 337–63 in W. Callebaut and R. Pinxten (1987).

Vollmer, G.
1984 "Mesocosm and Objective Knowledge." Pp. 69–121 in F. M. Wuketits (ed.), *Concepts and Approaches in Evolutionary Epistemology.* Dordrecht: D. Reidel.
1985 *Was können wir wissen?* Vol. 1: *Die Natur der Erkenntnis.* Stuttgart: S. Hirzel.
1986 *Was können wir wissen?* Vol. 2: *Die Erkenntnis der Natur.* Stuttgart: S. Hirzel.
1987a *Evolutionäre Erkenntnistheorie.* Stuttgart: S. Hirzel.
1987b "What Evolutionary Epistemology Is Not." Pp. 203–21 in W. Callebaut and R. Pinxten (1987).

White, L. A.
1956 "The Locus of Mathematical Reality: An Anthropological Footnote." Pp. 2348–64 in J. R. Newman (ed.), *The World of Mathematics,* vol. 4. New York: Simon and Schuster.

Wigner, E. P.
1960 "The Unreasonable Effectiveness of Mathematics in the Natural Sciences". *Communications in Pure and Applied Mathematics.* 13:1–14.

Wilder, R. L.
1981 *Mathematics as a Cultural System.* Oxford: Pergamon.

Wittgenstein, L.
1983 *Tractatus Logico-Philosophicus.* London: Routledge and Kegan Paul.

PART III: MATHEMATICS, POLITICS, AND PEDAGOGY

6

Mathematics as a Means
and as a System

The claims of sciences to influence what humans do vary. On the one side we have sciences like physics or sociology that are satisfied in describing what is. On the other side we have technical sciences or pedagogy, which give, more or less, hints on how humans should act.

One main issue of this book is that of the relationship between mathematics and society. Usually this topic is studied within disciplines like sociology or history of science. This chapter and the following ones have been written with the intention of *influencing* the relationship mathematics ↔ society through mathematics education. The corresponding academic discipline, didactics of mathematics, is in my view the collective effort to study *and to shape* the relationship between humans, on the one hand, and mathematics, on the other (cf. Fischer and Malle, 1985). Here I mean humans as individuals as well as collective social systems, or even as whole societies. I include of course, learning at the individual level, but also learning in the society, especially the structuring of the society by mathematics—as it is done, for example, through economics.

THE DUALITY OF MEANS AND SYSTEM

One of the fundamental ideas I argue for in this chapter is that mathematics provides a *means* for individuals to explain and control complex situations of the natural and of the artificial environment and to communicate about those situations. On the other hand, mathematics is a *system* of concepts, algorithms and rules, *embodied in us*, in our thinking and doing; we are subject to this system, it determines parts of our identity. This system runs from everyday quantifications to elaborated patterns of natural phenomena to complex mechanisms of the modern economy. On the basis of mathematical considerations, we define economical relations among people (who is indebted to whom and how much) and define what justice is. I, therefore, view mathematics on the one side as a means, which we can handle like a tool, and on the other side as a system, which we have to obey and which is inseparably connected with our social organization.

This does not mean that this system has always existed: by creating the means we developed the system and in the process of developing the system—in which organized education played an important role—new means became necessary. In other words, the aspect of means and the aspect of system are inseparable. Therefore I speak of a *duality of mathematics as a means and as a system*. It expresses also the fact that humans are subjects and objects of mathematics: we create means, which build up into a system and react back on us.

For example, the duality of means and system explains, in my view, the relationship between mathematics and the computerization of the world. Mathematics is in two respects a precondition for the process of computerization. As a means, it is a prerequisite for the technology. The computer is the continuation of the process of the materialization of abstract relations, for which mathematics always has been responsible: from calculating stones and abacuses up to the written symbols that have dominated mathematics in the last three centuries (cf. Fischer and Malle, 1985, chap. 6). As a system it predisposes humans to accept the computer, to handle it. This predisposition includes thinking in symbols and thinking using machine-like processes, but also a causal, logical, mechanized organization of social life, especially of labor (cf. Bammé et al., 1983).

My thesis is that we *have not yet learned to cope with the duality* of means and system in mathematics, we have especially not learned to recognize the reciprocal actions between these two aspects of mathematics. As a rule the "systemic" side is not under consideration or it is assumed without question. We do not know which conditions are behind the system and how it is changed by new means. Or the systemic side is criticized unreflectedly, for example, as a method of disciplining according to the interests of the ruling class (as in socialistic critical pedagogy). I think we should study the duality of means and system with more effort and try to handle it—politics of science, education, and so forth—more constructively.

VALUE OF MEANING AND VALUE OF UTILIZATION

By neglecting the systemic side of mathematics, we make the aspect of means absolute. Today the relevance of mathematics for society is primarily viewed in terms of its role as an efficient means for solving problems. This is also the case for the relevance of natural sciences and, in a weaker sense, also for that of other sciences. This is not the only possible way to explain the relevance of sciences for society, as historical studies show.

Sociologist Friedrich Tenbruck in his article "Der Fortschritt der Wissenschaften als Trivialisierungsprozeß" distinguishes between the "value of meaning" and the "value of utilization" of a scientific proposition. The former refers to the "content of meaning which it [a scientific proposition] can possess before and independent of its utilization" (Tenbruck, 1975:23; my translation). "Content of meaning" corresponds to the possibility of gaining orientations for society, for its philosophies of living. Value of meaning and value of utilization are not solely determined by the content of a scientific proposition. They depend on societal conditions, on already existing knowledge, and so forth.

Now Tenbruck formulates a "Law of Trivialization" (see fig. 6–1) as follows:

In the progress of knowledge the facts or laws lose their meaning. At the beginning they have a high value of meaning, but usually no value of utilization. At the end they have no value of meaning, but usually a high value of utilization . . . The progress of science provides more and more knowledge, but destroys their meaning . . . The process of trivialization shortens sciences down to facta bruta, propositions about mere facts. Science is no more a source of legitimation for society, or it becomes a very problematic source. (Tenbruck, 1975:23–24; my translation)

In his description of the process of trivialization in the natural sciences, Tenbruck claims that the value of meaning of the facts of natural sciences today is very low. Propositions about cosmology, ele-

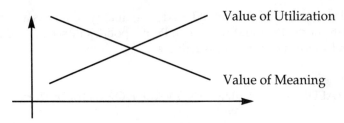

Figure 6–1.
The Law of Trivialization.

mentary particles, and so on run through our heads but do not really touch our thinking. Tenbruck states that even if physicists would go back to a geocentric system of the universe, this news would have value for only a few days. He writes:

> A revolution in physics is no longer a revolution in the heads of people. Whether nature is one way or another need not be of interest for us, as long as it is certain nothing can be discovered in nature but mere laws. This knowledge [that nature functions according to certain laws] each child sucks with his mother's milk and no scientific progress can add anything to it. (Tenbruck, 1975:24; translation mine)

This has not always been so, Tenbruck claims, and he presents a series of quotations from the history of natural sciences in the seventeenth, eighteenth, and nineteenth centuries. Scientists at the beginning of the modern natural sciences were interested in understanding a *message of God*. The laws of Newton have been viewed as the "last act of divine revelation" by which "God reveals the order in his creation." Later, with the increasing *secularization of science,* God lost his active role, and the belief that humans could by their own competence and using scientific methods solve the riddles of the universe became important. As a symbol of this belief one can take the famous answer of Laplace to Napoleon's question about the role of God in his system: "Sire, I have no need of that hypothesis."

But still there existed a high value of meaning in the sense of Tenbruck. The essential place of trivialization is located in the nineteenth century, when *scientific insight proved insufficient for understanding the world* and, above all, fundamentally insufficient for the "transformation of knowledge into the right human order" (Tenbruck, 1975:31; my translation). Tenbruck writes: "With this natural sciences left the competition for satisfying the needs of the people, but not without claiming importance via its orienting methodology in new disciplines, the social sciences, which had to fulfill the task" (Tenbruck, 1975:33; translation mine). This summarizes Tenbruck's thesis. One can argue against Tenbruck that he uses "value of meaning" in a narrow sense, because instrumental utilization has meaning too. Therefore I would rather use the term *value of promise* instead of value of meaning.

Another objection to Tenbruck's argument could be that the value of meaning and the corresponding orientations were accessible only to a small group of people. This group was indeed larger than that of the scientists. Tenbruck reports that in the seventeenth century, scientific topics were dealt with in sermons and newspapers, as well as in saloons and aristocratic circles; ladies were taught algebra, physics, geometry, and so forth. Nevertheless, most people did not come into contact with these things. But, the small group of leading persons in society included "opinion leaders."

The process of trivialization has taken place even in the heads of the scientists. Their motivation has changed. They do not believe in the promises of science as their colleagues in the seventeenth and eighteenth centuries did. Today scientists often are more alienated from their subjects.

It is clear that the value of utilization corresponds to the aspect of means in the means-system duality, but it is not the case that the value of meaning—or promise—corresponds to the aspect of system. The latter refers in a more comprehensive and sounder way to the connection between knowledge and society. But one can say, following Michael Otte, that the value of meaning (or promise) corresponds to the conscious part of the aspect of system. In the course of trivialization, this part vanishes, and the systemic part changes permanently, since new means arise, and, if this becomes evident, a new change in the value of meaning is possible. I will come back to this idea later on.

PROMISE AND REALITY IN MATHEMATICS

When, during the last three centuries, did the value of meaning of mathematics become clear? My suggestion: It was in the late eighteenth century and the nineteenth century, when mathematics was viewed as a universal method for explaining and manipulating the world, going beyond the range of natural sciences. I think of the many trials to describe social or economical situations with the help of mathematics. I think, for example, of the "physique sociale" of Adolphe Quetelet (cf. Porter, 1985) or of the development of statistics in Great Britain in connection with "Eugenics" (Mackenzie, 1981). Further, I think of the efforts in mathematizing thinking (logic) or language (mainly in this century). Simultaneously, the separation of mathematics from the natural sciences occurred.

Compared with the high promises at the beginning of the just mentioned efforts *the value of promise of mathematics has very much decreased.* Economists today concede that mathematical laws are of restricted value in their field and that forecasting is problematic. Sociologists and psychologists withdraw from a scientific paradigm, primarily oriented towards mathematical methods (Fischer, 1987a,chap. 3). The same is true for linguistics and other fields. On the other hand it seems that the value of utilization is still increasing, especially in fields outside the natural sciences. How is this contradiction possible?

I suppose that two things are important about this. First, a "modelling paradigm" of the applications of mathematics has been developed. This means that no tight connection between mathematics and "reality" is postulated; one does not expect that mathematically invented propositions are fully correct in reality. Secondly, mathematics often is not used to describe reality, but to *construct a new reality.* For this it is excellent. A good example is technology, where mathematics comes in via the natural sciences. In a more direct way, mathematical constructions are used in developing computers (mathematical logic) or in developing programming languages. Another field of examples for the constructive power of mathematics is economics. Economists invent rules to be obeyed when money is borrowed or stocks are sold; they construct indices that govern financial politics (taxes, supports, salaries, and so on). Of course these rules and indices sometimes are intended to describe reality, but in any case they create a reality and make negotiations and decisions possible.

Within such a constructed reality, mathematics can produce knowledge about this very reality. Mathematical optimization methods in business administration are the more successful the more mathematical structure is given—by laws, rules, and so forth. One of the most successful applications of statistics, quality control, presumes a production process running according to mathematical rules. We have a *circularity:* The more mathematics is used to construct a reality, the better it can be applied to describe and handle exactly that reality.

INSTRUMENTATION: OUTDISTANCING, AND
MIRRORING HUMANITY

As already mentioned, an increase of the value of utilization corresponds to an increased stress on the means aspect of the means-system duality. Let us consider this process under the modelling paradigm and in terms of the view that mathematics creates reality. We use mathematics for explaining and controlling something *different from us*. For this purpose we create mathematical constructs and put them as a "means" between ourselves and "the other" to be explained or controlled. Even if the other is not really different from ourselves, if it is the economy or the society, which we are parts of, or even nature, to which we humans belong, it becomes something other, different; it becomes an *object* to be studied or shaped. The more this *outdistancing* succeeds, the better mathematics functions. The more this outdistancing succeeds, the less mathematical knowledge has to do with ourselves and the less we can learn from it about ourselves. If meaning in the sense of Tenbruck is connected with self-recognition, then decreasing meaning is a consequence of instrumentation (putting means in between). The process of outdistancing is shown in figure 6–2.

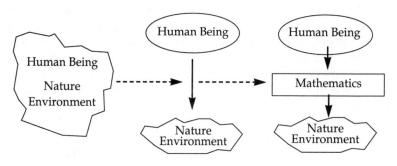

Figure 6–2.
The Process of Outdistancing.

To emphasize the system aspect and the reciprocal actions between means and system implies studying the relationship between humans and mathematics and recognizing that means act upon the thoughts, feelings, and actions of people. Some work has been done in this direction. Herbert Pietschmann (1980) has analyzed how mathematical-logical principles form the modes of how we recognize nature. Gerhard Schwarz (1985) has established a connection between these very principles and the main mode of social organization in our society, the hierarchy. Arno Bammé (1986) has pointed to several connections between recent ideas in the natural sciences and ideas in the social sciences. (For all this, see Fischer 1987a).

What does this mean with respect to Tenbruck's process of trivialization? I think that a *new value of meaning* could arise if the system aspect is stressed as sketched above, namely: *mathematics as a mirror of humanity*. It is possible to learn new facts about our ways of recognizing, thinking, and organizing by studying the system aspect of mathematics. The same is true for natural sciences and other sciences that have developed a high degree of instrumentalization.

If we compare this possibility with the original promises of modern natural sciences, we see, of course, a difference. The issue is no longer the absolute principles that were prior to and independent of humans, but the question, which assumptions, aims and (unconscious) decisions stand behind what we call science today?

SYSTEM AND MEANS IN MATHEMATICAL ECONOMICS

To illustrate the program sketched above I will use the case of mathematical economics. I think the relationship between mathematics and economics is especially direct and the system aspect especially evident; both constructivity (for example, constructing indices) and reflexivity (constructed laws acting back on ourselves) are obvious. Here, ignoring the systems-means duality has fatal consequences.

In his book *Anti-Equilibrium*, mathematical economist Janos Kornai classifies economics in a way which corresponds in some sense to the means-system duality. He distinguishes between the "real-scientific" part and the "decision-scientific" part (Kornai, 1975:10–13; my translation). The task of the former is the proper description of economical "reality," whereas the question for the latter is how, from this description, rules can be deduced for economical decisions. Mathematics is used primarily in the second area—this corresponds to a dominance of the means-aspect—though mathematical concepts and considerations of consistency can be of use also in the real-scientific part. The difference lies in the criterion of truth: in the first case it is coincidence with "reality"; in the second it is logical argumentation, because only consequences are deduced from given conditions (paradigmatic example: techniques of optimization). It seems plausible to adjoin the system aspect to the real-scientific part and the means aspect to the decision-scientific part.

Kornai complains about the lack of reflection or the intermixture of the real-scientific part. A good description of reality must exist, one that is independent from the decision-scientific part, before it is possible to think about optimal decisions. The prevailing intermixture, according to Kornai, consists in the fact that the basis of economical theory is a concept of 'rational economic man' with regard to which the individuals in the economy are supposed to make their decisions.

But, in a certain way, this intermixture seems inevitable. It is clear that motives and aims of people must be incorporated into the real-scientific theory. Further, and this is important for the role of mathematics, motives and aims in the economy are not independent from the means that are used in the economy and with the help of which decisions are made. Of course one can assume a global aim, "welfare," but in asking what this welfare is or when subgoals are defined (for example, to possess much money [means], to have as few workers as possible in a factory [in order not to pay too much in taxes or salaries]) things already become dependent on the means. And therefore the means, which often are expressions of collective motives, act upon the system. A very special example in today's economy is a means, "computer," which leads to a restructuring of the economy, not only because of new possibilities of material production, but also because of new situations for decisions. In general, one can conjecture that the intermixture of real-scientific and decision-scientific parts of economics is a consequence of the means-system duality in mathematics.

This does not mean that Kornai's critique is wrong. In my opinion, it is necessary to have a sound look at the interactions between means and system. In particular, one has to pose the question, which collective motives, assumptions, and goals have been incorporated into the economy (and economics)? In the following sections, I use a critical analysis of the "general equilibrium theory" as carried out by Kornai (1975:16–30).

FUNDAMENTAL ASSUMPTIONS OF THE GENERAL EQUILIBRIUM THEORY

A fundamental assumption of the so-called general equilibrium theory is the *separation of the economical world into consumers and producers*, at least with respect to a certain commodity. For each part there exists a theory that treats it as if it were independent of the other part; that means a theory of its behavior with given boundary conditions. More concretely: *The consumer tries to maximize utility.* Commodities are offered at different prices. The consumer selects in a way that maximizes his/her "utility function," the arguments of which are the quantities of the commodities. This problem of maximizing is solved for different assumptions of prices and this yields a *demand function*, which adjoins to each price a certain commodity demand. *The producer tries to maximize profit.* The producer's decisions concern how much of a commodity to produce at a given price. Again, solving this problem for each price yields a function, the *supply function*, indicating how much the producer is willing to offer at a given price. Now demand and supply meet at the *market*, and under certain circumstances an *equilibrium* arises. The main parameter is the *prices*, and it is—in the idealized model—the only content of the information processed among the market participants. They are anonymous to each other; the only interesting information is the price.

The theory provides that under certain assumptions (for example, "complete competition" of many market participants) the equilibrium is optimal with regard to "social welfare." But this is an optimality in a very weak sense. Each change in the situation would mean that there is at least one market participant whose situation becomes worse (Pareto optimality). This means, for example, if the sum of commodities to be distributed is constant, then each distribution is optimal. It is obvious that in some cases many people are not interested in such an "optimal equilibrium."

In summary, the basic elements of the mathematical theory of the economy are:

- it separates reality into subunits,
- the relations between these subunits are defined by the flow of money, commodities, and price information,
- Each subunit maximizes a certain function

The last activity is part of what economists call "rational acting." The meaning of rationality is described in the next section.

THE PREFERENCE ORDERING OF "HOMO OECONOMICUS"

A well-known theory of economical rationality starts with the assumption that each economically acting individual defines a so-called *preference ordering* on the set of possible actions. This is a relation ">" with two arguments, $a > b$ meaning that the alternative a is preferred to the alternative b. A preference ordering is at least irreflexive and asymmetric; that means

$$\forall x : \neg\ x > x \text{ and } \forall x, y : x > y \Rightarrow \neg\ y > x,$$

and sometimes, additionally, transitivity, that is,

$$\forall x, y, z : x > y \land y > z \Rightarrow x > z,$$

is postulated or serves as a criterion for the consistency of the economical thinking of the individual. Economist Amartya Sen has developed a fundamentally critical view of the concept of rationality as contained in the theory of preference orderings. His first point is that of a "circularity" in the idea of the "Homo oeconomicus," that human who always acts according to personal interests, expressed by individual preference orderings:

> The reduction of man to a self-seeking animal depends in this approach on careful definition. If you are observed to choose x rejecting y, you are declared to have "revealed" a preference for x over y. Your personal utility is then defined as simply a numerical representation of this "preference", assigning a higher utility to a "preferred" alternative. With this set of definitions you can hardly escape maximizing your own utility, except through inconsistency. Of course, if you choose x and reject y on one occasion and then promptly proceed to do the exact opposite, you can prevent the revealed preference theorist from assigning a preference ordering to you, thereby restraining him from stamping a utility function on you which you must be seen to be maximizing. He will then have to conclude that either you are inconsistent or your preferences are changing. You can frustrate the revealed preference theorist through more sophisticated inconsistencies as well. But if you are consistent, then no matter whether you are a single-minded egoist or a raving altruist or a class-conscious militant, you will appear to be maximizing your own utility in this enchanted world of definitions . . . This approach of definitional egoism sometimes goes under the name of rational choice, and it involves nothing other than internal consistency. (Sen, 1982:88–89)

This "definitional egoism," the only regulator of which is consistency, equated in the theory with the transitivity of the preference relation, can hardly be tested empirically:

If today you were to poll economists of different schools, you would almost certainly find the coexistence of beliefs (i) that the rational behaviour theory is unfalsifiable, (ii) that it is falsifiable and so far unfalsified, and (iii) that it is falsifiable and indeed patently false. (Sen, 1982:91)

What are not considered by this egoistical concept of preferences, but are according to Sen, meaningful for practical acting, are *commitments to various social groups*. Traditionally it is assumed that there is an irreconcilable gap between utility for the individual and utility for all. Sen argues that this contradiction neglects essential aspects of social life, especially the fact that there are social groups *between* the individual on the one side, and "all" (the society, the state) on the other side. Social groups can be decisive for the economical behavior of the individual: family, town, village, or region where he/she lives, a group of business friends, a union, and so on.

Furthermore, between the claims of oneself and the claims of all lie the claims of a variety of groups—for example, families, friends, local communities, peer groups, and economic and social classes. The concepts of family responsibility, business ethics, class consciousness, and so on, relate to these intermediate areas of concern, and the dismissal of utilitarianism as a descriptive theory of behaviour does not leave us with egoism as the only alternative. (Sen, 1982:85)

Sen concedes that perhaps the idea of maximizing individual utility may suffice for explaining private consumption. By no means is this the case, according to Sen, if *public goods* (such as streets, parks, the illumination of public places, and so forth) are concerned; they are used by many persons. A further example is that of the behavior of people during public elections (Sen, 1982:96–97). Nobody would go to elections if everybody decided according to individual utility, since it is not worth the trouble. The absurdity of the concept of maximizing individual utility is also demonstrated by the following joke:

"Where is the railway station?" he asks me. "There", I say, pointing at the post office, "and would you please post this letter for me on the way?" "Yes", he says, determined to open the envelope and check whether it contains something valuable. (Sen, 1982:96)

The critique of the concept of 'rational economical man', the base of some economical theories, formulated so far is: first, it is circular and not really testable and, second, it is wrong.

DIVISION OF RATIONALITY

If the concept of the rational economical man were correct, humanity would be in a fatal situation, since certain situations could not be coped with. This point is important, since a theory of social behavior with the property that, if reality behaves according to this theory this would lead to catastrophes, is bad, at least if the theory provides no alternatives.

Situations that cannot be coped with if people behave according to the concept of the rational economical man are those that need a *common interest* that is more than the sum of individual interests. A classical example for such a situation is provided by game theory, namely *"the prisoner's dilemma."* Two prisoners have commited a serious crime, but can only be convicted of a minor one. The district attorney separates them into two cells and tries to get them to confess. He offers to both the following bargain: if both confess, they must stay in prison for ten years. If both do not confess, each of them gets two years. If one confesses and the other does not, the confessing prisoner gets off scot-free whereas the other one gets twenty years. The situation is represented in table 6–1 (the first number in each field is the

		Prisoner 2	
		Confesses	Does Not Confess
Prisoner 1	Confesses	−10, −10	0, −20
	Does Not Confess	−20, 0	−2, −2

Table 6–1.
The Prisoner's Dilemma.

punishment for the first, the second for the second prisoner). What shall the prisoners do? Each one knows also the possibilities of the other. Sen writes:

Each Prisoner sees that it is definitely in his interest to confess no matter what the other does. If the other confesses, then by confessing himself this prisoner reduces his own sentence from twenty years to ten. If the other does not confess, then by confessing he himself goes free rather than getting a two year sentence. So each prisoner feels that no matter what the other does it is always better for him to confess. So both of them do confess guided by rational self-interest, and each goes to prison for ten years. If, however, neither had confessed, both would have been in prison only for two years each. Rational choice would seem to cost each person eight additional years in prison (Sen, 1982:63).

The prisoner's dilemma may seem esoteric, but we can establish connections between the dilemma and everyday life. Sen presents the following example. Assume the interests of a community prevent people from *throwing away empty bottles*; they must bring them back to the stores (because of pollution problems). What is the situation for the individual? The (pollution) harm I have to reckon with when I throw away bottles is perhaps less than the trouble I have to take to return the bottles. Of course I am interested in the fact that many people return their bottles. My preference ordering may look as follows: All people but me return their bottles; all people, including me, return their bottles; nobody returns bottles; only I return bottles. The situation can be represented in a table like that of the prisoner's delemma with only one number in each field, denoting the payoff for the individual (table 6–2).

		The Others	
		throw away	give back
I	throw away	3	1
	give back	4	2

Table 6–2.
The Bottle Dilemma.

If everyone in a community has this preference relation in mind, the result must be: nobody returns bottles. Whatever the individual assumes that the others will do improves the situation if the bottles are thrown away. (One aspect of the above argumentation is that the "individual" and "the others" are treated like equal, independent persons.)

The concept of 'Homo oeconomicus' and the theory of preference orderings correspond to a division of a possibly (hopefully) existing common rationality into rationalities of individuals. According to Sen, this theory describes the "rational fool," who acts consistently, but with narrow horizon, recognizing nothing outside his/her sphere; in particular, he/she is not able to think solidly, and thereby, in the end, acts against his/her own interests.

MACHINE VERSUS CREATIVE SOCIAL PROCESSES

One consequence of the preceding criticism of traditional mathematical economics is that it views the economy as a *machine* rather than as a living system. Of course it is not a "pure" machine; it does not function deterministically. There are degrees of freedom for the subunits, especially for the individuals, in whom free will is admitted. For instance, the consumers can choose their preference orderings. But the whole system has the following property of a machine: *it consists of well-defined parts, and the relationships between them are determined precisely.* In the present case these relationships are ruled by the flow of money and commodities and the exchange of price information. The matter of concern is only the relationships between the parts, and these are of interest only insofar as they provide "inputs" into these relationships. What is going on inside one part is of interest only insofar as it has an effect on the economical interaction with other parts. Obviously economists have the hypothesis that by abstracting from the (living) inner processes in the parts and by concentrating on the relationships between the parts one can get something invariant (dead). Life is restricted to the parts; its influence on the system as a whole is very limited. In this sense, the economy is viewed as a machine.

Further, in this view, a part does not and does not need to think about others or the whole. Nobody has to think about the whole, it simply functions.

Opposed to this machinelike view is the fact that the economy evidently is a living, creative social process. The creativity is expressed in the development of new technologies, new modes of communication (for example, via sales promotion); it lies in the emergence of new needs and of a new consciousness of society (*Zeitgeist*). With this *the relationships permanently change*, new groups arise, old groups vanish, and so forth; shortly, the "machine" becomes another one. This is not allowed for machines in the above sense: the separation into parts and the relationships would have to remain constant.

But also a second property of machines in the above sense is not fulfilled in living social systems: that parts do not care for the whole, that they are principally "less" than the whole. Living social systems, by contrast, have the property *that the whole can be mirrored in the parts*, which, via the relationships with other parts, can be images of the whole. And vice versa: the whole can be an image of the inside situation of parts, especially individuals. Thus it is plausible that subunits can change the whole system—as a rule starting from a position of opposing the system.

The two aspects of the "real" economy that are not considered by traditional mathematical economics, namely the dynamics of relationships and the possiblity that individuals are not less than the whole, are in a certain sense contained in the proposals of Kornai and Sen to improve the theory. Kornai (1975) developed a system-theoretic concept of the economy, where flows of information (not only price information) play an essential role. This implies an enrichment of the structure of relationships. What I find problematic even in this concept is that information is treated like a (material) commodity, which can be transferred from one (economical) unit to another, can be exchanged, and so on. This is not a comprehensive concept of information; such a comprehensive concept would include the changeing of relationships, the restructuring of the partition into subunits, and the emergence of a new consciousness or of collective will. Sen developed a concept of 'meta-orderings' of 'preference-orderings' according to a degree of morality. Morality can be viewed as an aspect of representing the whole (society) in the parts (the individuals). What is not considered is the process of communication by which morality is created. (By the way, the two ideas of Kornai and Sen, new possibilities of communication and high individual mortality, are those by which the prisoner's dilemma can be overcome.)

The question of the *relationship between the whole and its parts* seems to be crucial. Traditional economics assigns free will only to the individual (or other "atoms") and defines the whole system by a rigid partition into subunits and rigid relationships. Such a simple distinction between the whole and its parts is not adequate for social systems. But perhaps it is rooted in today's understanding of the fundamentals of mathematics, as I show next.

MATHEMATICAL ECONOMICS AND SET THEORY

In a lecture about the history of the relationship between mathematics and economics, Thomas Kuczynski (1987) conjectured that in the future mathematics will be essentially influenced by disciplines different from physics, especially by biology and economics. But in this he sees problems which demand a fundamental reconsideration of the basis of mathematics.

First, let us formulate it in the sense of Heisenberg: "Each particle consists of all the other particles" (Heisenberg 1967, p. 30). the behavior of each single particle therefore is determined by the behaviors of all the other 10^{80} particles . . . The particles therefore do not form a set in the sense of classical set theory, for according to Carnap's "Zirkelfehlerprinzip", a totality is not a set if the elements of the totality are determined by the totality itself. [Carnap 1931:98]

The proposition that one elementary particle is in reality the "ensemble" of cosmic relations was stated in an analogous mode 120 years before Heisenberg with respect to another kind of elementary particle: Karl Marx (1958: 6) says in his sixth Feuerbach thesis: "In its reality [the individual] is an ensemble of social relationships." The only difference lies in the fact that the elementary particle of physics does not mind if we neglect it in our considerations of the pico-or mega-cosmos, whereas the [human] individual permanently feels treaded on, especially if it is outside the three-sigma-interval of a statistical distribution . . . The fact that in models of mathematical economics we always isolate parts of the economy (we cut them away from others) is one of the fundamental problems, because we know only very little about the—practical and theoretical—admissibility of such a procedure. For instance, the critique that in problems of optimization too little of the restrictions and boundary conditions are considered in the model (environmental problems, social consequences, sociopsychological preferences, etc.) has its root in the fact that real economical systems are not sets in the sense of classical mathematics. (Kuczynski, 1987:160–61 my translation)

The main point is: *set theory forbids defining elements in terms of their relationships with all the other elements.* That sounds strange if one

remembers modern mathematics, where "wholes" are defined ax-
iomatically precisely in terms of determining the relationships between the elements. Think of a definition of natural numbers with the
Peano-axioms or of an axiomatic definition of a vector space, a group,
a ring, and so on. It is not common today to begin with a definition of
single numbers or vectors or group elements, and so forth. (This hint
is due to Fritz Schweiger.) But if you have a closer look at these axiomatic definitions, for example, the definition of a vector space,
which starts with "Let V be a set . . . ," you see that a "naked set" is
presumed. One assumes that as the basis of each structure there exists
a set without structure. And in this set the elements preexist without
relationships with the other elements.

The problem becomes still clearer if one seeks *constructive* definitions of (for example, algebraic) structures or representations of structures. Then one has to define the elements in a way such that the
structure is mirrored in them. But they never do this completely. Take
the Von Neumann definition of natural numbers: each natural number is the set of its predecessors, but no natural number mirrors the
structure of all natural numbers, not even of a subset, if it belongs to
it. In general, there always exists an unbridgeable gap between what
structure can be created inside one element and the whole structure
and, therefore, between hierarchically arising construction on the one
hand and global axiomatics on the other.

Of course even in social systems that total structure is not a priori contained in each element (individual), but there is such a potentiality. In other words, *the elements are able to learn about the whole system and their role in it.* In mathematical structures on the basis of set
theory there are very narrow bounds for this "learning process."
What is to be added further in social systems is that the elements can
develop a *consciousness* about their relations with other elements and
a *will* and thereby gain the possibility of changing the relationships
and thereby the total structure.

One can say that mathematics would be fundamentally overly
criticized if one would postulate that it should depict reality in its
models in such a way that the inseparability of parts and whole and
the dynamics of relations are taken account of. One can say that mathematics has to separate and has to hold constant, partially with the
aim of negotiating about what is to be held constant. But then one
should know this and keep in one's mind that there is a principal difference between mathematical models and living social reality, which
cannot be overcome by abstraction toward the level of relationships.
And that the dynamics of social systems possibly results in perma-

nently overcoming mathematizations and similar systems—whereby I think of mathematizations via technology as well as those which have a more direct impact on social systems.

REFERENCES

Bammé, A.
1986 "Wenn aus Chaos Ordnung wird: Die Herausforderung der Sozial-wissenschaften durch Naturwissenschaftler." University of Klagenfurt. Manuscript.

Bammé, A., G. Feurerstein, R. Genth, E. Holling, R. Kahle, and P. Kempin
1983 *Maschinen-Menschen, Menschen-Maschinen: Grundrisse einer sozialen Beziehung*. Reinbek: Rohwolt.

Burkhardt, H.
1981 *The Real World and Mathematics*. London: Blackie.

Carnap, R.
1931 "Die logizistische Grundlegung der Mathematik." In R. Carnap and H. Reichenbach (eds) *Erkenntnis*, vol. 2. Leipzig: Meiner

Dörfler, W., R. Fischer, and W. Peschek, eds.
1987 *Wirtschaftsmathematik in Beruf und Ausbildung*. Vorträge beim 5. Kärntner Symposium für Didaktik der Mathematik. Vienna: Hölder-Pichler-Tempsky; Stuttgart: B. G. Teubner.

Fischer, R.
1987a "Mathematik und gesellschaftlicher Wandel." (Mathematics and Social Change, Chapter 11, this volume). Projektbericht im Auftrag des Bundesministeriums für Wissenschaft und Forschung. University of Klagenfurt.

———.
1987b *Begrüßungsadresse: Einige Gedanken zum Tagungsthema*. In Dörfler, Fischer, and Peschek (1987).

Fischer, R., G. Malle, with H. Bürger
1985 *Mensch und Mathematik: Eine Einführung in didaktisches Denken und Handeln*. Mannheim: Bibliographisches Institut.

Heintel, P.
1979 "Thesen zu einer Philosophie der Mathematik." University of Klagenfurt. Manuscript.

Heisenberg, W.
1967 Einführung in die einheitliche Feldtheorie der Elementarteilchen. Stuttgart: Hirzel.

Kornai, J.
    1975   *Anti-Äquilibrium: Über die Theorien der Wirtschaftssysteme und die damit verbundenen Forschungsaufgaben.* Berlin, Heidelberg, New York: Springer.

Kuczynski, T.
    1987   Einige Überlegungen zur Entwicklung der Beziehungen zwischen Mathematik und Wirtschaft (unter besonderer Berücksichtigung der Entwicklung der Wirtschaftsmathematik). In Dörfler, Fischer, and Peschek (1987:145-66).

Mackenzie, D. A.
    1981   *Statistics in Britain 1865–1930: The Social Construction of Scientific Knowledge.* Edinburgh: Edinburgh University Press.

Marx, K.
    1958   *Thesen über Feuerback.* In K. Marx and F. Engels, *Werke,* vol. 3. Berlin: Dietz.

Needham, J.
    1956   "Mathematics and Science in China and the West." *Science & Society* 20:320–43.

Pietschmann, H.
    1980   *Das Ende de naturwissenschaftlichen Zeitalters.* Vienna, Hamburg: Zsolnay

Porter, T. M.
    1985   "The Mathematics of Society: Variation and Error in Quetelet's Statistics." In *British Journal of the History of Science* 18:51–69.

Restivo, S.
    1983   *The Social Relations of Physics, Mysticism, and Mathematics.* Dordrecht, Boston: D. Reidel

Roberts, F. S.
    1979   *Measurement: Theory with Applications to Decisionmaking, Utility and the Social Sciences.* Reading, Mass.: Addison-Wesley

Schwarz, G.
    1985   *Die "Heilige Ordnung der Männer": Patriarchalische Hierarchie und Gruppendynamik.* Opladen: Westdeutscher Verlag.

Sen, A.
    1982   *Choice, Welfare and Measurement.* Cambridge, Mass.: MIT Press.

Spengler, O.
    1973   *Der Untergand des Abendlandes.* Munich: Deutscher Taschenbuch Verlag.

Tenbruck, F. H.
1975 "Der Fortschritt der Wissenschaften als Trivialisierungsprozeß." N. Stehr and R. König (eds.), *Wissenschaftssoziologie: Kölner Zeitschrift für Soziologie und Sozialpsychologie*, 18:14–47.

7

Reflections on the Foundations of Research on Women and Mathematics

"Women and mathematics" has been an object of systematic scientific analysis, especially in psychology, pedagogy, and mathematics didactics, for about three decades now. During the last few years the extent of the research has increased. This is at least partially due to the raised public awareness of gender-related issues in general. The level of research activity is especially high in Anglo-American countries.

The lower participation rate of females in mathematics in comparison with males is considered as the main problem and therefore is the essential motive for dealing with the topic of "women and mathematics." At least in Western industrial nations investigations show again and again that fewer females than males enroll in noncompulsory mathematics courses or study mathematics at the university (Armstrong, 1985; Chipman and Thomas, 1985; Metz-Glöckel, 1987). Because school systems in Anglo-American countries allow students to select subjects for study, the student can withdraw from courses, and from mathematics in particular. Having this in mind, the research interest in these countries in women and mathematics becomes clear. There is a secondary problem, too—the minor mathematics achievement of females occurs only in certain contexts: Boys outdo girls in higher grades in solving specific tasks of high cognitive complexity (Fennema, 1980; Hanna, 1988).

The general aim of scientific research on gender-related aspects of teaching and learning mathematics is to find out the causes of the stated female deficiencies. The findings, it is assumed, will lead to a reliable basis of knowledge for the development of intervention models that can do away with the deficiencies. Several programs have already been created at some schools and universities (cf. Kaiser-Meßmer, 1989). Although the goal in this research is to get to the bottom of the deficiencies, not every empirical study establishes correlations between achievement and, respectively, enrollment in mathematics and influencing factors; quite a few studies concentrate on a detailed analysis of the underlying factors themselves. Essential findings about intrapersonal factors of the learners (Boswell, 1985; Fennema and Sherman, 1977; Joffe and Foxman, 1988; Pedro et al., 1981) are that

- mathematics is considered a male domain, and mathematics is stereotyped as male more by boys than by girls
- boys have a more positive attitude towards mathematics than girls in some respects (for example, boys consider mathematics as more useful for their future occupations than girls do)
- boys show greater confidence in their ability to learn mathematics than girls do.

The major outcome of the studies dealing with teacher-student interactions in the mathematics classroom is that in nearly every observed category teacher-boys interactions are more frequent than teacher-girls interactions (Becker, 1981; Eccles and Blumenfeld, 1985; Öhler, 1989; Stallings, 1979).

I do not want to outline the numerous findings in this field; I would, rather, like to describe the background of thinking the research in general is based on. I do not deny that there are studies that look at the topic from other perspectives, start from other theoretical approaches, and use methods other than those mentioned below (Burton, 1986; Jungwirth, 1990a, 1990b; Maines, 1985; Walkerdine, 1989). Most of the research, however, is settled in a relatively homogeneous conceptual framework. The aim of this chapter is to make this framework clear by describing some of its main aspects. I want to show its limits as well as its potentiality, and I want to try to demonstrate that the common background is just one possible basis one can start from. It should be remarked in advance that the underlying conceptual framework is not often described explicitly in papers. It is the empirical studies that predominate. The framework has to be reconstructed mainly by examining the design of a study (which sort of phenomenon is analyzed by which sort of methods), or by considering the introduction (which publications are mentioned or quoted), or by reading thoroughly the final discussion (which interpretations of the data are given). Because of this, the appropriate method to elucidate the background of thinking is the "method of documentary interpretation," as it is presented by Wilson:

> Dokumentarische Interpretation besteht darin, daß ein Muster identifiziert wird, das einer Reihe van Erscheinungen augrude liegt; dabei wird jede einzelne Erscheinung als auf dieses zugrunde liegende Muster bezogen angesehen—als ein Ausdruck, als ein "Dokument" des zugrunde liegenden Musters. Dieses wiederum wird identifiziert durch seine knokreten individuellen Erscheinungen, so daß die das Muster wiedergebenden Erscheinungen und das Muster selbst einander wechselseitig determinieren.

> Documentary interpretation means that a pattern is identified which is at the root of a number of phenomena. Thereby each phenomenon is considered as related to the underlying pattern—as an expression, as a "document" of this pattern. This, on the other hand, is identified by the various concrete phenomena so that the phenomena and the pattern itself determine each other. (Wilson, 1980:60; my translation)

PREMISES ABOUT TECHNOLOGICAL PROGRESS AND GENDER DIFFERENCES

The very fact that women's relation to mathematics and not men's relation is the center of interest is informative. It is actually not so obvious that there is a problem with the relations of females and mathematics. This makes sense only if certain presuppositions are made. Firstly, the "technological formation" (Hülsmann, 1985) has to be considered as natural; that is, we assume that modern technology will spread over all sectors of labor and spheres of life and therefore women too will be forced to turn towards mathematics, science, and engineering in order not to lose their chances in the future labor market and to find their way in the future society. The interest in the research on women's relation to mathematics is also explained by this argument. From a point of view, however, that doubts whether technological progress is in fact progress, males' turning to mathematics would be considered a problem whereas females' distance would be a starting point from which to begin reviewing technological development in general. There is a second presupposition, too, a presupposition with regard to gender differences. The common argument shows that the framework is *Gleichheitsdiskurs*, "equity discourse" (Prengel, 1986). Gender difference is seen as a hierarchy with the male on the top and the female on the bottom: the attitudes, behavior, life-style, and so forth of males are at least implicitly much more appreciated than those of females. The male is the standard; and, applying this standard, all female-specific activity seems to be deficient. Equity is understood as an approximation to the standard set by males. But from a point of view that puts a higher value on the life forms of females than on those of males (the "radical-feminist discourse"; Prenge. 1986) the distance of women from mathematics would not be a problem and a failure that has to be criticized.

THE EXPLANATION MODEL: SOCIALIZATION

One important aspect in the background of thinking about women and mathematics is the model of how the specific relation of women to mathematics comes about. There are two fundamental positions that can be characterized by the catchwords *nature* and *nurture*. Within the research community there is a broad consensus that women's relation to mathematics is a problem for which a simple explanation will not do. It is assumed that there is a large number of influencing factors. Biological ones are not denied, but there is a clear tendency to favor social ones (Hanna and Leder, 1990; Schildkamp-Kündiger, 1982). Lower participation rates and achievement on the part of the females are considered to be a result of gender-specific socialization, as, for example, the following quotation shows:

> The basic assumptions underlying our research program are: (1) that sociocultural factors are transmitted primarily through parents, peers, and the educational settings; (2) that these factors profoundly shape women's attitudes toward mathematics; and (3) that attitudes subsequently affect women's performance in mathematics. This type of model is consistent with the views of current constructivist theorists (e.g. Harvey, Hunt & Schroder, 1961) who assert that individuals construct their experience in accordance with their beliefs about reality. Many of these beliefs

relate to what American society traditionally has deemed appropriate (or inappropriate) as roles for women. These cultural mores and dictums are transmitted in the form of sex-role stereotypes. As a consequence of accepting these stereotypes, women are directed away from mathematical pursuits. (Boswell, 1985:175–76)

This position implies the following research program: "One must examine how these social conditions affect the educational environment of a female as well as affect the personal belief system of each learner. Both the educational environment and what a person believes about her/himself has a direct influence on what is learned in mathematics" (Fennema, Walberg, and Merrett, 1985:304). Maybe the tendency to refrain from biological explanations is grounded in the assumption that attributing gender-specific differences to nature would label them as unchangeable. This is not compatible with the aim to equalize females' and males' relation to mathematics. "Due to nature," does not, however, mean unchangeable in every case. There are different conceptions of how the stated deficiencies come about by socialization. These conceptions can be attached to the three models of the acquisition of gender-specific behavior in general (Schenk, 1979).

The first conception says that the relation to mathematics comes about by imitation and identification. There is a large number of role models (for example, parents, teachers, or figures in stories) who show certain relations to mathematics that the children imitate. This conception is especially evident in statements indicating the lack of appropriate role models for girls because of the small number of female mathematicians, or the differences in the presentations of females and males in mathematics textbooks (females seem to be less mathematically competent than males; cf. Eckelt et al. 1987; Glötzner, 1982).

A second conception says that girls and boys are treated differently in the context of learning mathematics and therefore develop different relations to mathematics. This conception refers to the thesis of "differential socialization." Accordingly, the different treatment of the socialization agents is decisive for the development of gender-specific attitudes and modes of behavior. For example, this assumption is made in studies that analyze the behavior of teachers towards girls and boys in the mathematics classroom.

These two models show the behaviorist approach in the research on women and mathematics. The third model—which is the predominating one—stems from cognitive approaches in social psychology. It says that a child develops a concept of which kind of relation to mathematics would be appropriate to her or his gender; and this concept orients the child's behavior towards mathematics. According to this model, girls and boys do not simply react to certain stimuli but digest them on the basis of cognitive structures establishing meaning and reducing complexity. In this model the two following explanations play an important role.

One explanation of the lower mathematical achievement of females refers to the "fear of success" construct. Accordingly, the fear of negative consequences of the attainment of success in male domains, such as unpopularity or doubt about their femininity, excludes females from high achievement. Since mathematics is stereotyped as male, this is the case too when females solve mathematical tasks (Leder, 1985).

Another model refers to attribution (Weiner, 1974). Accordingly, the attributions by which an individual explains her or his own success or failure influences her or his future achievement behavior. Attributing success to abilities and failure to bad luck or less effort increases the future achievement expectations, whereas doing it the other way around—success is due to effort and luck, failure is due to a lack of ability—decreases the expectations. There is some indication that the first attribution pattern is characteristic for males and the latter for females, including cases where they solve mathematical tasks (Leder, 1981; Wolleat et al., 1980).

GENDER-ROLE EXPECTATIONS ARE THE DETERMINING FACTOR

The gender-role stereotypes are considered as essential schemata of orientation within the socialization processes that affect typical male and female relations to mathematics. It is assumed that the stereotypes cause women to avoid mathematics. They give girls the "wrong" models, they lead to "wrong" expectations on the part of the parents, teachers, and so forth and to their "wrong" treatment of the girls, and they make the girls themselves think that a (successful) occupation in mathematics contradicts femininity. More or less, it is assumed that females (and males) are compelled to comply with the gender-role expectations. They apparently have no choice. This assumption shows a certain affinity of the background of thinking to the "normative paradigm" of sociology (Wilson, 1980). Within this paradigm, interactive behavior is determined by the role expectations individuals are exposed to. The stability of interaction structures is explained by the hypothesis that an individual feels the need to behave in the way she or he is expected to. Here it is the relation to mathematics that is explained by this model. Research following the cognitive approaches follows this model too. This is not surprising because, as Stangl (1989:179) says, "es wird in der kognitiven Psychologie zwar auch die Aktivität des Individuums bei der Informationsverarbeitung betont und so zu einem eher passiven Rezeptionsmodell des Behaviorismus abgegrenzt, doch bleiben die postulierten innerpsychischen Prozesse letztlich einem linearen oder konditionalen Kausalitätsmodell verpflichtet" ("in cognitive psychology the active part the individual plays in information processing is emphasized and thereby separated from the model of passive reception typical for behaviorism but nevertheless the postulated intrapsychical processes are explained by a linear or conditional model of cause and effect"; my translation). The claim is that socialization, in terms of the gender-role stereotypes, causes women to avoid mathematics.

Females get a specific status: they seem to be the victims of the gender-role expectations determining behavior in general and toward mathematics in particular. For example, Leder looks at the matter in this way when she says that there are, in comparison with biological constraints, "far greater pressures imposed by social and cultural stereotypes about cognitive skills and occupations"(1985:305). The idea that there is a role-making aspect instead of only role taking (Turner, 1962), that is, that an individual interprets and shapes her or his role, is not discussed. From such a point of view the actual actions of an individual could turn out to be those that are the most reasonable in the given contexts from her or his perspective. References to the idea that women themselves actively establish their relation to mathematics in the prevailing form cannot be reconstructed. The critical psychologists point out the problem of how victims (or objects) of acting can change into autonomous active subjects of acting. In her reasoning about the milieu-theoretical approaches in feminist research, Haug (1980:89) hits the sore point of this model: "Dabei geschieht dann die Verwandlung derer, die bislang als bloße Opfer von Behinderung erschienen, in Subjekts ihres Handelns wie im Märchen, wenn der Frosch zum Prinzen wird" ("The transformation in the behavior of subjects who up to now have seemed to be mere victims of discriminatory socialization is like the process in fairy tales when the frog becomes a prince"; my translation).

GENDER-ROLE STEREOTYPES LIMIT THE RESEARCH HORIZON

Interpretations of empirical results given or accepted by researchers show a tendency to remain within the framework built by gender-role stereotypes. It seems that gaining insights is limited by these stereotypes.

An example is sustaining the common interpretation of the construct "fear of success" which serves as an explanation for the lower mathematical achievement of females. This construct was developed within the interpretation of responses of women to the Thematic Apperception Test. When reviewing the responses to the test, Sassen (1980) found out that there is an alternative meaning of these responses; they do not need to be considered expressions of fear of success. They show the anxiety arising in women who find themselves unable to give meaning to the situation presented in the test. This inability is explained by the assumption that the test cue illustrates competitive success, which females have not integrated in their structure of knowing: "Their structure of knowing is more oriented toward preserving and fostering relationships than toward winning . . . Women bring a relational, contextual structure of knowing to the cue they are asked to make sense of, and thus find they cannot accommodate it to this kind of competitive success" (Sassen, 1980:19).

With this in mind, it seems that women do not have fear of success in mathematics but are confronted with analogous problems of making sense out of the situations where the data about their attitudes towards mathematics and their mathematical achievement are gathered. Their more negative responses could show just such problems. It might be that this is not taken into account because the category "sense" or "subjective meaning" does not play any role in defining the female and the male stereotypes.

Apart from this specific question, no attention has been paid either to the category "subjective meaning" in the analysis of females' relation to mathematics. Sassen's interpretation suggests, however, that it should be taken into account. The question about the subjective meaning of mathematics might turn out to be a central point within the discussion of gender differences in relation to mathematics. Maybe there are different approaches to mathematics, different ways to experience mathematics as meaningful, and different aspects of mathematics that make or do not make sense; and maybe mainly those ways are ignored and those aspects are emphasized in the mathematics classroom that make females rather than males feel uncomfortable because of a lack of meaning. To turn away from mathematics in this view would be a decision of females who make fruitless efforts to create meaning out of what is going on. They would not be victims of gender-role expectations; they would like to live in worlds that make sense.

CAUSES ARE LOCATED INSIDE THE INDIVIDUAL

The specific female relation to mathematics is considered to be due to certain conditions in the individual. This is a feature that all the concepts have in common. Intrapersonal characteristics such as attitudes, motives, cognitions—in particular, internalized gender-role stereotypes—are assumed to cause the behavior. As Ulich (1976:47) states: "Das Individuum erscheint als abstract-isolierter Träger konstanter Eigenschaften, die ihn zu bestimmten Verhaltensweisen 'befäfhigen' oder 'bewegen' (motivieren)" ("The individual appears to be an abstract-isolated possessor of constant attributes which enable or move her or him to behave in a specific way"; my translation).

The possibility that attributes of individuals are socially constituted and therefore connected with the social situation the attributes developed in, is not taken into account. For example, the mathematical competence of a student is assumed to be the result of her or his task solving only—she or he gives correct answers, makes sensible suggestions, or asks intelligent questions. Mathematical competence is not considered as a product of the social interaction in the classroom, that is, it is not considered as an attribute that is constituted by certain practices all participants, teacher and students, use to master the situation (Jungwirth, 1991b).

Those studies that analyze interactions between the teacher and the students in the mathematics classroom are also based on the assumption that regularities are caused by constant attributes in the individuals. It is typical for such studies to analyze the behavior of the teacher isolated from the behavior of the students. Implicitly it is supposed that the events in an interaction are the "sum" of the behaviors of the individual participants, the "sum" of the factors determining the behavior. The interaction is only the social place in which the true determinants operate.

In this type of thinking, the conception that the participants in an interaction interpret specific actions, as well as the whole situation, and design their own actions on the basis of these interpretations is missing. The habit of using the expression "behavior" instead of "acting" shows this: theoretical approaches based on the hypothesis that an acting person constructs a certain meaning of her or his action and that the addressed person attaches her or his subjective meaning to it, as in phenomenological sociology, symbolic interactionism, and ethnomethdology (Blumer, 1969; Cicourel, 1973; Garfinkel, 1967: Goffman, 1959; Schütz, 1962–66) use the expression "acting." Without this hypothesis it cannot come to mind that the participants in an interaction, by mutually adjusting their actions, can achieve results that cannot be traced back to attributes in the individuals. For example, it is outside the horizon of these studies that the failure of a student can become a little mistake (as it could be reconstructed in teacher-boys interactions) or a serious lack of understanding (as it could be reconstructed in teacher-girls interactions) according to the reactions of the student to the teacher's stating the incorrectness of the solution, the actions the teacher feels compelled to take in response to the student's reactions, the following answers of the student, and so forth (Jungwirth, 1991a).

THE NOMOLOGICAL-DETERMINISTIC PARADIGM

The research on women and mathematics is, in general (social) psychological research within the "nomological-deterministic paradigm" (Stangl, 1989:101). In this paradigm, it is the aim of psychology to discover laws and to explain and predict behavior through those laws (Stangl, 1989). As Ulich (1976:29) says, psychology wants "Invarianzen zu erforschen, die man in der Über-Zeitlichkeit und Unveränderbarkeit von Merkmalen, Ereignissen und Beziehungen zu entdecken hofft" ("to investigate invariable accounts that one hopes to find out in the timelessness and unchangeableness of attributes, events, and relations"; my translation). Easily, science and its methodology are recognized to be the norm. Therefore, it is clear that the empirical analyses are designed as verifications and falsifications of hypotheses by the methods of statistics and that the procedures follow the criteria of objectivity, reliability, and validity. The epistemological position is that scientific knowledge mirrors or has to mirror reality. From a constructivist point of view, on the contrary, concepts, models, theories, and so on would be considered as viable constructions that "überleben, solange sie die Zwecke erfüllen, denen sie dienen, solange sie uns mehr oder weniger zuverlässig zu dem verhelfen, was wir wollen" (survive as long as they serve their purposes, as long as they help us, more or less reliably, to get what we want; von Glasersfeld, 1987:141; my translation).

In view of the demand to change the status quo, the nomological-deterministic conception of the observed events, that is, the conception that human acting is strictly determined, leads to the same fundamental logical problems as are described by C. Kraiker, referring to behaviorism and psychoanalysis:

> Gleichzeitig entwickeln sie jedoch Anleitungen zur Änderung des menschlichen Verhaltens, sie geben Anweisungen, wie durch bestimmte Maßnahmen Handeln und Verhalten modifiziert werden können. Das Problem, das Rätsel bzw. puzzle besteht nun darin, daß hier zwei Dinge behauptet werden: Erstens, die Welt läuft nach unabänderlichen Gesetzen ab; zweitens, man kann dieses und jenes tun, damit die Welt sich ändert.

(Simultaneously they make suggestions for changing human behavior; they give instructions about how acting and behavior can be modified by specific interventions. The problem or puzzle is that two contradictory statements are made: first, the way things go is determined by unalterable laws; second, one can do this or that in order to change things. (Kraiker, 1980:172, quoted in Stangle, 1989:101; my translation).

In this framework, to change things means to make use of the laws as far as possible to improve the present state. With reference to the current topic, this means using the laws to improve the relation of women to mathematics by developing intervention models. It depends on the particular structure of such a program whether the conception of man customary in psychology (Stangl, 1989:29)—"*der machbare Mensch*," "the man who can be molded" (my translation)—will be the prevailing idea too or whether this conception is no longer valid and consequently the program centers rather upon the subjective views and ideas of the participating women.

Summing up, it can be said that the claim to find out the causes of women's specific relation to mathematics, the intention to change it, the psychological approach to the problem and the use of empirical-analytical methods correspond to each other. The background of thinking seems to have a certain "logic" of its own. This is the basis of scientific analyses of "women and mathematics" that seek to achieve the objectives I have noted. Maybe the relation to mathematics is predestined to be looked at from the given perspective because of mathematics' assumed status as an objective, absolute body of knowledge in general. It is striking that mathematics itself is not questioned in the empirical studies.

These reflections on the background of thinking in research on women and mathematics cannot be attained within its boundaries. If you are interested in the subjective meaning females and males attach to mathematics and their ways of making mathematics their own (that is, if you want to emphasize the contextuality of their relations to mathematics), or if—more generally—you prefer a systemic-ecological approach to the topic and rather want to ponder over problems than to solve them, then the predominating approaches and models will not have the qualities needed.

REFERENCES

Armstrong, J. M.
   1982 "Correlates and Predictors of Women's Mathematics Participation." *Journal for Research in Mathematics Education* 13, 2 :99–109.

   1985 "A national assessment of participation and achievement of women in mathematics." Pp. 59–95 in S. F. Chipman, L. R. Brushiand, D. M. Wilson (eds.), *Women and Mathematics: Balancing the Equation*. Hillsdale. N.J.: Lawrence Erlbaum.

Becker, J. R.
   1981 "Differential Treatment of Females and Males in Mathematics Classes." *Journal for Research in Mathematics Education* 12, 1:40–53.

Blumer, H.
   1969 *Symbolic Interactionism: Perspective and Method*. Englewood Cliffs, N. J.: Prentice-Hall.

Boswell, S. L.
   1985 "The Influence of Sex-Role Stereotyping on Women's Attitudes and Achievement in Mathematics." Pp. 175–98 in S. F. Chipman, L. R. Brush, and D. M. Wilson (eds.), *Women and Mathematics: Balancing the Equation*. Hillsdale N.J.: Lawrence Erlbaum.

Burton, L.
   1986 *"Femmes et mathématiques: y-a-il une intersection?"* Paper presented at the conference "Femmes et mathématiques," June 6, 1986, Quebec, Canada.

Chipman, S. F., and V. Thomas
   1985 "Women's Participation in Mathemtics: Outlining the problem." Pp. 1–24 in S. F. Chipman, L. R. Brush and D. M. Wilson, (eds.) *Women and Mathematics: Balancing the Equation*. Hillsdale N.J.: Lawrence Erlbaum.

Cicourel, A. V.
   1973 *Cognitive Sociology: Language and Meaning in Social Interaction*. Harmondsworth: Penguin

146   *Mathematics, Politics, and Pedagogy*

Eccles, J. S., and P. Blumenfeld
1985 "Classroom Experiences and Student Gender: Are There Differences And Do They Matter?" Pp. 79–114 in L. C. Wilkinson and C. B. Marret (eds.), *Gender Influences in Classroom Interaction*. New York: Academic Press.

Eckelt, I., G. Effe-Stumpf, and U. Geuenich-Brackly "Frauen und Mathematik:
1987 Mädchen im Mathematikunterricht nicht mehr unterbuttern." *Mathematikklehren*, 25:48–53.

Fennema, E.
1980 "Sex-related Differences in Mathematics Achievement: Where and Why." Pp. 76–90 in L. H. Fox, L. Brody, and D. Tobin (eds.), *Women and Mathematical Mystique*. Baltimore: Johns Hopkins University Press.

Fennema, E., and J. A. Sherman
1977 "Sex-related Differences in Mathematics Achievement, Spatial Visualization and Affective Factors." *American Educational Research Journal* 14, 7:51–71.

Fennema, E., H. Walberg, and Cora Merrett "Introduction to: 'Explaining Sex-
1985 related Differences in Mathematics: Theoretical Models'." *Educational Studies in Mathematics* 16, 3:303–5.

Garfinkel, H.
1967 *Studies in Ethnomethodology*. Englewood Cliffs, N.J.: Prentice-Hall.

Glasersfeld, E. von
1987 *Wissen, Sprache und Wirklichkeit: Arbeiten zum Radikalen Konstruktivismus*. Braunschweig and Wiesbaden: Vieweg.

Glötzner, J.
1982 "Heidi häkelt Quadrate, Thomas erklärt die Multiplikation: Rollenklischees in neuen Mathematikbüchern." Pp. 154–58 in Ilse Brehmer (ed.), *Sexismus in der Schule*. Weinheim and Basel: Beltz.

Goffman, E.
1959 *The Presentation of Self in Everyday Life*. Garden City, N.Y.: Doubleday.

Hanna, G.
1988 "Mathematics Achievement of Boys and Girls: An International Perspective." In D. Ellis (ed.), *Math & Girls*. Ontario Educational Research Council

Hanna, G. and G. Leder
1990 "International Conference—The Mathematics Education of Women." International Organization of Women in Mathematics Education *Newsletter* 6, 1:4–11

Haug, F.
1980 *Frauenformen, Alltagsgeschichten und Entwurf einer Theorie weiblicher Sozialisation.* Berlin: Argument-Verlag, Sonderband 45

Hoffman, L.
1988 "Mädchen/Frauen und Naturwissenschaft/Technik." Pp. 2–15 in S. Giesche and D. Sachse (eds.), *Frauen verändern Lernen.* Dokumenation der 6. Fachtagung der AG Frauen und Schule. Kiel: Hypatia.

Hülsmann, H.
1985 *Die technologische Formation—oder: lasset uns Menschen machen.* Berlin: Europäische Perspektiven.

Joffe, L. and D. Foxman
1988 *Attitudes and Gender Differences, Mathematics at Age 11 and 15.* National Foundation for Educational Research, Windsor, Great Britain.

Jungwirth, H.
1991a "Die Dimension 'Geschlecht' in den Interaktionen des Mathematikunterrichts." *Journal für Mathematikdidaktik* 2/3:133–171
1991b "Zur Konstitution mathematischer Kompetenz in der unterrichtlichen Interaktion." Pp. 133–137 in Beiträge zum Mathematikunterricht. Vorträge auf der 24. Bundestagung für Didaktik der Mathematik vom 26. 2. bis 2. 3. 1991 in Salzburg. Bad Salzdetfurth: Franzbecker.

Kaiser-Meßmer, G.
1989 "Frau und Mathematik—ein verdrängtes Thema der Mathematikdidaktik." *Zentralblatt für Didaktik der Mathematik* 2, 56–67.

Leder, G.
1981 "Learned Helplessness in the Classroom?" Pp. 192–303 In J. P. Baxter and A. T. Larken (eds.), *Research in Mathematics Education in Australia,* vol. 2. Kelvin Grove: Queensland.
1985 "Sex-Related Differences in Mathematics: An Overview." *Educational Studies in Mathematics* 16, 3:314–18.

Metz-Glöckel, S.
1987 "Locht und Schatten der Koedukation." *Zeitschrift für Pädagogik* 4:455–75

Öhler, J.
1989 "Interaktionen im Mathematikunterricht in geschlechtsspezifischer Hinsicht." Diploma thesis. Kiel:University of Kiel.

Pedro, J. D., P. Wolleat, E. Fennema, and A. Becker "Election of High School Mathematics by Females and Males: Attributions and Attitudes." *American Educational Research Journal* 18:207–18.

Prengel, A.
1986 "Konzeptionelle Planung der Untersuchungen des Feministischen Interdisziplinären Forschungsinstituts." Pp. 21–54 in Hessisches Institut für Bildungsplanung und Schulentwicklung (ed.), *Konzept zum Vorhaben Verwirklichung der Gleichstellung von Schülerinnen und Lehrerinnen an hessischen Schulen.* Wiesbaden.

Sassen, G.
1980 "Success Anxiety in Women: A Constructivist Interpretation of its Sources and its Significance." *Harvard Educational Review* 50, 1:13–25.

Schenk, H.
1979 *Geschlechtsrollenwandel und Sexismus: Zur Sozialpsychologie geschlechtsspezifischen Verhaltens.* Weinheim and Basel: Beltz

Schildkamp-Kündiger, E.
1982 An international review of gender and mathematics. ERIC Clearinghouse for Science, Mathematics, and Environmental Education, in Cooperation with Center For Science and Mathematics Education. Columbus, Ohio State University.

Schütz, A.
1962–1966 *Collected Papers.* The Hague: M. Nijhoff.

Stallings, J.
1979 "Factors influencing women's decisions to enroll in *SRI International advanced mathematics courses:* Executive summary." Final report. SRI International Menlo Park, Calif.

Stangl, W.
1989 *Das neue Paradigma der Psychologie: Die Psychologie im Diskurs des Radikalen Konstruktivismus.* Braunschweig and Wiesbaden; Vieweg.

Turner, R.
1962 "Role-Taking: Process Versus Conformity." Pp. 20–40 in A. M. Rose (ed.) *Human Behavior and Social Process: An Interactionist Approach.*

Ulich, D.
1976 *Pädagogische Interaktion: Theorien erzieherischen Handlns und sozialen Lernens.* Weinheim and Basel: Beltz.

Walkerdin, V. and the Girls and Mathematics Unit. *Counting Girls Out.* London: Virago Press.
1989

Weiner, B.
1974 *Achievement Motivation and Attribution Theory.* Morristown, N.J.: General Learning Press.

Wilson, T. P.
1980 "Theorien der Interaktion und Modelle soziologischer Erklärung." Pp. 54–80 in Arbeitsgruppe Bielefelder Soziologen (ed.), *Alltagswis-*

sen, Interaktion und gesellschaftliche Wirklichkeit, Opladen: West-deutscher Verlag.

Wolleat, P., J. Pedro, A. Becker, and E. Fennema
1980 "Students' Causal Attributions of Performance in Mathematics."
Journal of Research in Mathematics Education 11:356–66.

8

Politicizing the Mathematics Classroom

When an association between mathematics and politics is mentioned, educators usually think of informal logic, graphs, and statistics as topics that are useful in civics and that might be taught in mathematics classes. A familiar approach is to consider what subject matter the two areas have in common and to suggest interdisciplinary units or lessons that may contribute to understanding in both areas.

There is, however, another way in which to look at the association. In order to participate intelligently in political life, people need both practice and motivation. Neither is adequately considered in most systems of education. Even though many of us have discarded the notion that people can become instant participants in political life at age eighteen or twenty-one if they have adequately learned the history and structure of their own government, our school systems still function as if this sort of preparation were sufficient.

In this chapter, I want to suggest that classrooms, including mathematics classrooms, should be politicized; that is, students should be involved in planning, challenging, negotiating, and evaluating the work that they do in learning mathematics. Students who are actively involved in steering their own educations may become citizens who feel more capable of influencing civic affairs. Further, such involvement may also contribute to the learning of mathematics. The discussion will proceed in three parts: first, I will discuss the paradox of "privilege and oppression" that characterizes student life and consider how Paulo Freire's pedagogy might be applied to mathematics classrooms: second, I will discuss the same problem from a feminist perspective and make a few suggestions in line with this mode of thinking; finally, I will look briefly at the current constructivist movement and argue that it must be embedded in an ethical or political framework if it is to be effective in reforming classroom practice.

PRIVILEGE AND OPPRESSION

Students are, from most perspectives, privileged persons, and the longer they are students the more privileged they become with respect to both present and future status. "When I was your age," parents often say to sons and daughters, "I was already at work" (or supporting a family or bearing a child). Student status is considered to be, and on some grounds is properly judged to be, one of privilege. But there is another side to student life: Paulo Freire says, "Any situation in which 'A' objectively exploits 'B' or hinders his pursuit of self-affirmation as a responsible person is one of oppression" (Freire, 1970:40).

It can be argued that our usual ways of proceeding in mathematics classrooms do indeed hinder the development of our students as responsible persons. It is teachers who are said to have worthwhile knowledge, who set the tasks and the dates for their completion, who judge the results and control the conditions under which students work. As Freire describes it:

> Education thus becomes an act of depositing, in which the students are the depositories and the teacher is the depositor. Instead of communicating, the teacher issues communiques and makes deposits which the students patiently receive, memorize, and repeat. This is the "banking" concept of education, in which the scope of action allowed to the students extends only as far as receiving, filing, and storing deposits. (1970:58)

Educators often declare themselves to be innocent of Freire's charges on the grounds that they use inquiry methods of higher order questions or because they give challenging problems on tests and, therefore, do not expect mere "receiving, filing, and storing." But there is more to Freire's complaint than a challenge to our choice of content and its transmission and testing. Fully human beings have some control over the conditions in which they work, the evaluation of their own efforts, and the level of rewards for which they are willing to work. These are matters we rarely consider sharing with students, even though we are aware that such concerns are central to the motivation of adult citizens and continually debate ways to motivate students in mathematics. Modes of schooling that require students to give up control—or, worse, never even consider the possibility that they might exercise control—over the essential elements of their own development are, clearly, poor preparation for civic life in free societies.

John Dewey also wrote on this issue. Referring to Plato's definition of a slave as "one who executes the purposes of another" (1963:67), Dewey described students in traditional schools as slavelike in their subordination to teachers. His recommendation was that students be directly involved in the construction of the objectives that would guide their own learning. While this at first sounds reasonable, on closer inspection it has always seemed practically impossible for mathematics teachers. How can students be involved in the selection and construction of objectives when the material at hand is sequentially prescribed by its very nature and the students are as yet unaware of its possibilities?

Elementary schools that attempted to use Dewey's ideas often encouraged students to design and conduct projects, investigations, and experiments in which mathematical problems would inevitably present themselves. The mathematical instruction provided by teachers would, then, fit the needs of students who were already energetically pursuing ends both they and their teachers deemed valuable. It was never clear, however, that such methods could be applied effectively to secondary instruction in mathematics, and, indeed, Dewey's own writing on the progressive organization of subject matter (see Dewey, 1963:73–88) suggested that secondary school students might be ready for more traditional instruction in the sense that they would then understand its structure and give assent to its objectives. Thus secondary school mathematics was never greatly influenced by Dewey's suggestions.

Difficulties that arise in trying to apply the ideas of Freire and Dewey may be attributed to the emphasis both writers and their interpreters put on subject matter. Dewey, of course, urged a balance in emphasis between the demands of subject matter and the interests of students, but he advised teachers to study students interests that have a direct relation to the subject matter. In following Dewey on this, we tend to overlook the main interests of adolescents. We strive mightily—and often fruitlessly—for direct connections between adolescent life and formal mathematics.

Freire's discussion of "themes" has many similarities to Dewey's ideas on student interest and occupations. Both men are concerned with growth rather than specific fixed ends. Themes, for Freire (1970:92), are "generative" when "they contain the possibility of unfolding into again as many themes, which in their turn call for new tasks to be fulfilled" (1970:92). This unfolding opening up into new possibilities is, for Dewey, a mark of growth. Mathematics teachers, however, are baffled by the seemingly Herculean task of locating themes, basic interests, or occupations that might serve as springboards for the systematic study of algebra, geometry, or trigonometry.

My contention is that our focus on subject matter in this regard is a mistake. Mathematics teachers can prescribe the topics their students will study, and high school students will find this procedure for the most part unobjectionable. The themes teachers must attend to are developmental themes: Who am I? Who will I be? How hard should I work and toward what end? How am I doing and how can I tell? Can I make a difference in the world? Do I have any control over my own life?

When themes of this sort are considered, it becomes clear that the classroom must be politicized to explore them productively. Students need a voice in the conduct of their own education. Recognizing that teenagers are concerned about their present and future identities and why they should work at learning mathematics, teachers might discuss a range of possibilities with them. First, mathematics teachers should admit that people need different amounts of mathematics and that lots of talented people find mathematics difficult and even boring. As sighs of relief are heard all over the classroom, the teacher can open a dialogue that will lead to negotiation. A true dialogue, Freire reminds us, is open, that is, conclusions are not reached before the dialogue begins. The sort of discussion in which teachers try to justify their own ready-made decisions also has a place in classroom discourse, but it is not dialogue.

Let me describe one possible scenario that might be played out as teacher and students negotiate their work together. Let's say the course is Algebra Two-Trigonometry. The teacher describes a range of topic coverage that will allow him or her to certify that a student has completed a year's work in mathematics. What will I need to go on to precalculus? one student asks, and several others echo the concern. Teacher and students eventually define a standard course that will make this move possible. Another student complains that he doesn't plan to take any more math, but he needs this third year's credit for entrance into the state university. The teacher asks what the student plans to do professionally, and it is clear from his answer that mathematics is not vitally involved in that work. Several other students express similar interest in what will ultimately be negotiated as a minimum course. Then someone asks about the possibility of an enriched or honors–type course, and this is worked out.

The possibilities now are several. The teacher may provide three different syllabi, each designed to cover the negotiated number of chapters. Homework assignments may be differentiated. Different levels of chapter tests may be constructed. Working groups will have to be set up. Sometimes teachers decide to team teach in large sections so that the required flexibility can be maintained. Students may change their minds after getting started and decide that they want to do more or less than they originally planned. However teachers decide to handle their classes under this kind of arrangement, it is clear that they will do far less lecturing to whole groups, and students will have to assume some responsibility for teaching each other. When teachers explain to both students and parents that such group work has great advantages particularly for those who teach and for those who ask questions and get them answered (Webb, 1982), the proposed plan should be both politically and academically acceptable.

Besides learning mathematics, students in an appropriately politicized classroom are learning to take responsibility for their own academic progress. They should also be involved in their own evaluation and grading. Self-evaluation is an important element in what William Glasser (1990) calls the "quality school." A big part of a teacher's job is to help students develop appropriate criteria for self-evaluation. This means shared responsibility for evaluation. Teachers and students can negotiate what will be required to earn an A, B, C, or pass in the course. Actual contracts can be signed, or an overall agreement can be reached so that students know exactly what they have to accomplish each marking period. The teacher may want to use a modified form of mastery learning in arriving at both the variations in course structure and the grading scheme. In my own work with high-school students, I did exactly this. Students had to pass a given unit of work (usually a chapter) before proceeding to the next, and they had to pass a designated (negotiated) number of units to get a particular grade.

As we listen to students' themes, we become aware in a keenly personal way that they are concerned—even anxious—about their own capacities and potential for success. A system that reduces test anxiety by allowing tests to be retaken encourages students to discuss their weaknesses freely. They begin to evaluate their own readiness to take tests as they build a background of experience in passing/not passing tests. When the only penalty for not passing a test on the first try is a loss of time, students are encouraged to explore their own patterns of mastery. A very large benefit that may be gained from this arrangement is that students will believe them when teachers say that *learning* is the aim of the work they are doing. If students show that they have learned on the first try, fine; if not, the important thing is that they *do* learn, and the teacher is always on the students' side in this cooperative enterprise.

What I have been describing—in bare outline because of space constraints—is a variant of Keller Plan. Extensive research on this method has shown it to be especially effective in college mathematics (Kulik, Kulik, and Cohen, 1979; Dunkin, 1986), and greater use at the secondary level is likely to show similar results. As mathematics teachers, we want our students to achieve, of course, but the point I have been trying to make here is that achievement is not our only goal. We are involved, also, in the development of fully human beings—people who will be able to pursue self-affirmation responsibly. To pursue self-affirmation responsibly requires participation in cooperative political life. Keller Plans, modified to include group work, are illustrative of academically effective ways to politicize classrooms.

A FEMINIST NOTE

The classroom politics I have discussed so far involves negotiation and shared planning. This mode of politics does not focus narrowly on power, although forms of power are inevitably present. Some feminists have put enormous emphasis on power relations whenever we interact with other human beings. This view seems to overlook a host of other affects and purposes that characterize human relations, and further, as Jean Elshtain (1981) has pointed out, a term tends to lose all meaning when it is stretched to cover everything. It seems clear, however, that there is a political dimension to personal life, and I have been arguing similarly that there is a political dimension of classroom life.

When we look at classrooms from the perspective of Freire or Dewey, we see that there is no question of *whether* there is a political dimension; the proper question focuses on what sort of politics should be promoted. There is a long line of feminism—including such diverse figures as Virginia Woolf, Pearl Buck, Jane Addams, and Betty Friedan—that emphasizes a politics of empowerment compatible in most respects with that of Freire and Dewey. This school of feminism has sometimes been called "social feminism," but it might better be described as "transformational feminism." It seeks to transform the social world into a place where all human beings can pursue self-affirmation responsibly. Power is involved, of course. If we seek to empower, we use some of our own power to help others acquire the power they need to exercise some control over their own lives. The aim is not, however, a pursuit of power for its own sake or for the purpose of coercing people to do whatever the persons in power deem right and just. The aim is to distribute power so that all persons can attain a reasonable level of efficacy and so that greater attention can be given to other aspects of human life. When we empower our children to make responsible choices for certain details in their own daily lives, we have more time to enjoy their company or to pursue our own activities. Similarly, when we empower students to learn, to teach, to evaluate their own work (with our help), we free up time for enrichment and personal guidance.

From this perspective, we should not be so concerned with motivating everyone to do well in mathematics but, rather, with giving everyone a chance to find out whether he or she is interested in doing mathematics. To reject the study of mathematics as a free and well-informed decision is the choice of a responsible citizen; to plod through it docilely is a slavelike response, and to drop out without reflective consideration is to lose an opportunity both to learn mathematics and to learn about oneself. In a politicized classroom, students become citizens who have some control over their academic lives. This means promoting dialogue both within mathematics lessons and about mathematics as a potential avenue of self-affirmation.

One way to promote dialogue within mathematics lessons is to allow students to work together. The question, How shall we do this problem? becomes a genuine question when it is not asked by one who is presumed to have the answer. With proper training, students can learn to draw each other out, build on each other's suggestions, and express their appreciation for good ideas and hard work (Cohen, 1986). There is considerable promise in such procedures for both enhanced mathematics learning and preparation for cooperative political life.

Another way to promote dialogue within mathematics lessons is for the teacher to ask, How shall we do this problem? and then to follow student contributions to their logical conclusions. This sort of lesson cannot be conducted with every topic or on every day, because of time constraints, but it can profitably be used occasionally. A variant is to probe for alternative ways of solving problems or proving theorems, rather than pushing for one quick way to a single right answer.

Still another opportunity for dialogue within mathematics lessons arises when students encounter word problems. We are sometimes inclined to ignore the content of problems in the belief that it is appropriate to concentrate on mathematical structure. In the real world, however, people engage in mathematical activity to solve problems that involve objects, concepts, and events that are themselves of central interest. Some studies have shown that the content of word problems favors boys over girls (Chipman and Thomas, 1985; Donlon, Ekstrom, and Lockheed, 1976); that is, the objects and events included in typical word problems are more familiar to and of greater interest to boys. If this is so, we might, of course, vary the content to include matters of interest to girls, but we might also make it a point to discuss the content of problems so that students become acquainted with a wider range of mathematical applications and also with current research on gender differences.

Engaging in genuine dialogue within mathematics classrooms quite naturally leads to dialogue about mathematics as an area of personal interest. Girls especially might profit from such dialogue because there are still powerful stereotypes to be overcome concerning women and mathematics (Boswell, 1985). In an appropriately politicized classroom, both girls and boys should be able to discuss the ways in which they may eventually use mathematics professionally. There is a caution to be observed in this area, however. Girls should not be made to feel that their self-worth depends on their entering a field that requires mathematics. There is a long and dismal history of telling women that they should be more like men in order to be worthy. (In medieval days, for example, it was suggested that women could be "more like men" by remaining virgins and living in convents! This choice supposedly made a woman more worthy spiritually.) It is thus important to emphasize that both women and men may be interested in and competent in mathematics but that no one needs to excel in mathematics in order to be a valued member of society.

Finally a feminist perspective on mathematics classrooms suggests that some attention should be given to ways of knowing. In order to understand mathematics and to understand ourselves and others, we need to gain an appreciation for the various ways in which people come to know. Feminist scholars are now uncovering alternative modes of thinking in both ethical (Gilligan, 1982; Noddings, 1984) and epistemological (Belenky et al, 1986) contexts, and there seems to be a general increase of interest in this area (Eisner, 1985). Dialogue about mathematics and mathematical activity includes discussion about our own thinking—how it proceeds, what enhances it, how it may differ from that of others. Such dialogue belongs in a politicized mathematics classroom, and, as I have argued throughout this chapter it may well promote both mathematical achievement and effective participation in sociopolitical life.

A WORD ON CONSTRUCTIVISM

There is a movement in mathematics education that is growing in influence and popularity. it is called "constructivism." (See Davis, Maher, and Noddings, 1990.) I want to say a little about it here, because it offers some positive possibilities but, at the same time, contains the seeds of its own destruction. Unless it is embedded in an encompassing moral position on education, it risks categorization as a *method*, as something that will produce enhanced traditional results.

On the positive side, constructivism is a cognitive position holding that all mental acts, both perceptual and cognitive, are acts of construction. No mental act is a mere copy or externally imposed response. If you pass some information to me, I must actively listen to make out what you are saying, and then I have to fit what I have heard into what I already know and decide what to do with the new material. What I do with it depends on my purposes. I may respond with a sympathy-like interest because I care about you even though I do not care about the topic you are discussing. Or I may care deeply about the topic and realize that what you are saying suggests that I have been wrong about something. Or I may evaluate your remark and decide that *you* have made a mistake.

Constructivists believe that people are internally motivated and that they construct their own mental representations of situations, events, and conceptual structures. Constructivist teachers, then, usually spend time trying to find out what their students are trying to do and why. They are ready with suggestions and challenges that will help students to make strong and useful constructions.

What motivates constructivist mathematics educators? For the most part, they want to teach mathematics in ways that are compatible with their beliefs about how people learn. They also tend to believe that mathematical thinking is rich, complex, tolerant of ambiguity, filled with attempts that may or may not succeed, and broadly useful in many human activities. Believing all this, they want to promote their students' mathematical growth.

But constructivism as a pedagogical orientation has to be embedded in an ethical or political framework. The primary aim of mathematics teachers cannot be to promote mathematical growth, although that is certainly *one* worthy goal. Rather, the primary aim of every teacher must be to promote the growth of students as competent, caring, loving, and lovable people. Teachers with this aim will work flexibly in teaching mathematics—inspiring those who care about mathematics for itself to inquire ever more deeply, helping those who care instrumentally about mathematics to prepare for the line of work they desire, and supporting as best they can those students who wish they never had to encounter mathematics. To have uniformly high expectations for all students in mathematics is morally wrong and pedagogically disastrous. It is part of a sloganized attempt to make our schools look democratic and egalitarian when, in fact, they are systems continually struggling for tighter control.

Constructivists must attend to the fact that purposes are constructed as well as knowledge, and students have a large variety of purposes. Just as we need to know how students think if we are to help them build powerful mathematical constructions, we need to know them as persons if we are to assist their construction of well chosen purposes.

CONCLUSION

I have argued that mathematics classrooms should be politicized: Teachers should attend to the generative themes of adolescence and explore how mathematics is or is not involved in them; students should be involved in negotiating their own workloads and working conditions; students should participate in the evaluation of their own work; feminist themes of empowerment should be discussed, and mathematical thinking should be examined from these perspectives; finally, constructivist mathematics educators should consider the ethical and political dimensions of learning mathematics as well as the cognitive aspects.

REFERENCES

Belenky, M. F., B. M. Clinchy, N. R. Goldberger, and J. M. Tarule
1986 *Women's Ways of Knowing.* New York: Basic Books.

Boswell, S. L.
1985 "The Influence of Sex-Role Stereotyping on Women's Attitudes and Achievement in Mathematics." Pp. 175–198 in S. F. Chipman, L. R. Brush, and D. M. Wilson (eds.), *Women and Mathematics: Balancing the Equation.* Hillsdale, N.J.: Lawrence Erlbaum.

Chipman, S. F., and V. G. Thomas
1985 "Women's Participation in Mathematics: Outlining the Problem." Pp. 1–24 in S. F. Chipman, L. R. Brush, and D. M. Wilson (eds.), *Women and Mathematics: Balancing the Equation.* Hillsdale, N.J.: Lawrence Erlbaum.

Cohen, E. G.
1986 *Designing Group Work.* New York: Teachers College Press.

Davis, R. B., C. Maher, and N. Noddings, eds.
1990 *Constructivist Views on the Teaching and Learning of Mathematics.* Reston, V.: National Council of Teachers of Mathematics and the Journal for Research on Mathematics Education.

Dewey, J.
1963 *Experience and Education.* New York: Macmillan.

Donlon, T. F., R. B. Ekstrom, and M. Lockheed
1976 "Comparing the Sexes on Achievement Items of Varying Content." Paper presented at the annual meeting of the American Psychological Association, Washington, D.C.

Dunkin, M. J.
1986 "Research on Teaching in Higher Education." Pp. 376–91 in M. C. Wittrock (ed.), *Handbook of Research on Teaching.* New York: Macmillan.

Eisner, E. W., ed.
1985 *Learning and teaching the ways of knowing.* 84th Yearbook of the National Society for the Study of Education. Chicago: NSSE.

Elshtain, J. B.
1981 *Public Man, Private Woman.* Princeton: Princeton University Press.

Freire, P.
1970 *Pedagogy of the Oppressed.* New York: Herder & Herder.

Gilligan, C.
1982 *In a Different Voice.* Cambridge, Mass.: Harvard University Press.

Glasser, W.
1990 *The Quality School.* New York: Harper & Row.

Kulik, J. A., C-L.C. Kulik, and P. A. Cohen
1979 "A Meta-analysis of Outcome Studies of Keller's Personalized System of Instruction." *American Psychologist* 34:307–18.

Noddings, N.
    1984 *Caring: A Feminine Approach to Ethics and Moral Education.* Berkeley and Los Angeles: University of California Press.

Webb, N.
    1982 "Interaction and learning in small groups." *Review of Educational Research* 52:421–45.

9

The Dialogical Nature of Reflective Knowledge

INTRODUCTION

Mathematical education has political dimensions, but it is far from obvious how to point out these dimensions. Mathematical education could be conceived as part of a cultural conflict between Western countries and Third World countries, as in the writings of D'Ambrosio (1985). In a traditional setting it may result in a "cultural imperialism" conveying the essence of the rationalistic tradition integrated into the social and political structures of highly technological societies.

Mathematical education could also be interpreted as a method for stratifying students according to abilities that are established as socially important; at the same time, this method is conceived as objective, as it does not seem to incorporate personal sympathies and antipathies. No school subject has so well elaborated ways of testing. Testing methods in other subject areas have been critically examined, but testing in mathematics has not been critiqued to the same extent. Stratification based on mathematical abilities is socially accepted, although this acceptance may be quite illegitimate.

It is impossible to imagine the continued existence of a highly technological society, if technological progress and innovations do not continue. As an integrated part of technology, mathematics has important social and political functions, although, often, they may be difficult to identify within the classroom and in social practice. Too often it is impossible for the teacher to exemplify where to use the mathematics presented at the blackboard. Mathematics in today's society is made invisible; it is integrated into technological know-how.

If the implicit mathematics of society has to be made explicit as a step toward the identification of the political dimension of real applications of mathematics, mathematical education has a role to play, but a new role. To point out and to discuss the possible social functions of real applications of mathematics presupposes analysis related to mathematics, but not mathematical knowledge in a traditional context. How to develop this broader reflective competence is an open question.

Going into the classroom we also meet political aspects of mathematical education. Students gain or lose prestige according to their behavior patterns and communication styles. The rituals of communication in a mathematical classroom obey laws of a specific nature, which are not in accordance with communication patterns in other school subjects or in accordance with "natural" communication. The students have to play specific roles according to the "objective" stratification, implying that they have to develop specific learning strategies, as exemplified by the instrumentalist attitude discussed by Stieg Mellin-Olsen (1981).

A radical solution may be to ignore the whole subject; but it is the wrong conclusion to draw. It is possible that mathematical education has progressive as well as reactionary possibilities. A practical as well as a theoretical task must be to find out how to determine which ones are which. Is the goal to elaborate a curriculum involving information about use and misuse of mathematical modelling? Or is it to create open situations in which the students and the teacher may interact uninhibited by the standard rituals of communication? (This contradiciton is discussed in Skovsmose, 1990a.)

I do not maintain that the solution has to be found by means of theoretical reflection. Experimental practice may anticipate theoretical ideas. However, in what follows I have a more modest aim in sight: I want to identify some basic aspects of traditional epistemological theories that seem to blind us to what may be important if we want to react pedagogically to the political dimensions of mathematical education. I do not maintain that epistemological considerations must take priority. I choose epistemology only as my point of departure.

As I only want to identify a couple of characteristics of traditional epistemology, I simply hope to point out a couple of characteristics of an alternative epistemological approach. I wish to throw light on a form of educational practice to show that its claimed political neutrality is an illusion. However, it is not my intention in this context to try to analyze the specific content of the political dimension of mathematical education.

CLASSICAL MONOLOGICAL THEORIES

According to the rationalistic tradition in philosophy, knowledge is obtained by an individual act. The only thing important is to use *ratio* in deducing new knowledge from already obtained knowledge. Of course, the problem is where to start the deduction, and René Descartes has tried to exemplify a basic proposition (*Cogito, ergo sum.*), which needs no supporting argumentation. When you understand the proposition, you also understand that it must be true; and accompanying this understanding you have a criterion of knowledge. The rationalistic enterprise could then take the shape of a knowledge expansion by logical deduction.

Ratio constitutes a part of the individual, for which reason interpersonal relationships have no role to play in the development of knowledge. Knowledge development and communication are separated. In this sense we characterize the epistemological theory as *monological*. We may communicate knowledge, but communication is not a condition for obtaining knowledge. The genuine sources of knowledge constitute a part of the individual—although often ignored or not paid sufficient attention.

Monologism is also a distinct feature of classical empiricism. Obtaining knowledge is an individual process. The sources of knowledge are the senses, and so it is not necessary to interact with others. In *A System of Logic*, first published in 1843, John Stuart Mill discusses the nature of scientific knowledge and also the nature of mathematical knowledge. Mill belongs to the empiricist tradition, according to which concepts and ideas that are not related to sensation must be illusions. The fundamental question is, of course, if mathematics can be interpreted as an empirical science, or if it is of another nature.

Previously David Hume had formulated a thesis of duality in *Enquiry Concerning Human Understanding*. All objects of human reason may be divided into two kinds: matters of facts and relations of ideas. The first kind constitutes the natural sciences; the second, logic and mathematics. However, Mill wants to avoid this thesis of duality; instead he argues in favor of the idea that all sciences, including mathematics, have the same empirical nature.

Mill finds that mathematical propositions are not formal; they make references, and they have meanings. Numbers in the abstract do not exist. In this sense Mill asserts a nominalism: "Ten must mean ten bodies, or ten sounds, or ten beatings of the pulse. But though numbers must be numbers of something, they may be numbers of anything. Propositions, therefore, concerning numbers have the remarkable peculiarity that they are propositions concerning all things whatever; all objects, all existences of every kind, known to our experience" (Mill, 1970:167).

The content of a mathematical proposition concerns physical facts. The expression "two pebbles and one pebble" and "three pebbles" refer to the same aggregation of objects, but Mill underlines that they do not stand for the same physical fact. The two expressions have the same denotation but are different in connotation. This makes it possible for Mill to specify the content of an elementary mathematical proposition like 1+2=3: "it is a truth known to us by early and constant experience—an inductive truth: and such truths are a foundation of the science of numbers. Fundamental truths of that science all rest on the evidence of sense; they are proved by showing to our eyes and our fingers that any given number of objects, ten balls, for example, may by separation and rearrangement exhibit to our senses all the different sets of numbers the sum of which is equal to ten" (Mill, 1970:169).

The empiricist interpretation of mathematics has been criticized by Gottlob Frege and by the logicists in an effort to avoid the psychologism of Mill and to make logic the only basis for mathematics. But Mill's empiricism has been rehabilitated. Bloor (1976) in particular pays much attention to Mill's interpretation of mathematics, and it has great importance for mathematical education. Mill had already noted: "All the improved methods of teaching arithmetic to children proceed on a knowledge of this fact (that mathematical knowledge is empirical). All who wish to carry the child's mind along with them in learning arithmetic; all who wish to teach numbers, and not mere ciphers—now teach it through the evidence of the senses" (Mill, 1970:179).

Of course it is possible, according to empiricism, to misinterpret sense impressions; but such a misinterpretation is caused by psychological or sociologically mediated disturbances. If we are able to eliminate these "ideological" influences, we will be able to reach pure facts in the shape of sense data. True sense data are not mediated by any socially constructed system, such as language. It is possible for the individual to collect facts and from these facts to produce knowledge by induction.

I find it problematic that the experimental tradition in mathematical education often is accompanied by a monologism that focuses attention on the individual learner. Mathematical knowledge is assumed to grow from the individual basis of sense experiences, by some sorts of individual processes. I believe it is important to identify and criticize this monologism. Monologism is also entangled in modern interpretations of knowledge development.

MODERN MONOLOGICAL THEORIES

The genetic epistemology developed by Jean Piaget has played a central role in today's discussion of mathematical education. My thesis is that it too must be avoided if we want to outline the political character of mathematical education from an epistemological point of view.

It is Piaget's intention to relate both to the rationalist and to the empiricist tradition, without degenerating into one of them. Piaget makes a distinction between two different sorts of experiences: one has to do with the physical properties of objects and the other, with the operations that the subject may carry out with the objects. The distinction between the two experiences is fundamental to the identification of the genetic roots of pure mathematics.

Piaget maintains that mathematical knowledge is not created by an abstraction from physical properties of objects—which is the opposite of some empiricist interpretations of, for instance, geometric concepts: the concept of the plane is an abstraction from physical planes, and so on. However, Piaget does not maintain that mathematical concepts are without an empirical basis. His thesis is that mathematical knowledge is founded in the actions (operations) carried out on objects: "The essential fact is that . . . logico-mathematical experiences have to do with the actions which the subject carries out on the objects" (Beth and Piaget, 1966:232).

In *Genetic Epistemology*, Piaget summarizes his basic idea. He asks from what basis logical and mathematical knowledge is abstracted, and he continues: "There are two possibilities. The first is that when we act upon an object, our knowledge is derived from the object itself . . . But there is a second possibility: when we are acting upon an object, we can also take into account the action itself, or operation if you will . . . In this hypothesis the abstraction is drawn not from the object that is acted upon, but from the action itself. It seems to me that this is the basis of logic and mathematical abstraction" (Piaget, 1970:16).

Classical empiricism develops a "copy theory." The subject is passive; it receives impressions, and the impressions condense into concepts. Contrary to this, Piagetian epistemology stresses activity. The subject has to do something to obtain knowledge. Knowledge is created because of manipulation of objects.

However, one further step has to be taken, according to Piaget. The subject has to reflect upon the operations. He/she has to make reflective abstractions to obtain mathematical knowledge. One problem is that the operations with objects seem to be individual. Is it possible that this variety of operations could become the foundation of mathematics? All individual features have to be eliminated. Piaget makes a distinction between a "psychological subject" and an "epistemological subject." The first subject is individual, while the other is of a common nature and makes up the basis for mathematical knowledge. By virtue of "reflective abstractions" the subject makes implicit schemes of actions into explicit logico-mathematical patterns of thought.

This outline of the genetic epistemology makes obvious that it has a monological character. The operations of the subject do not involve any interpersonal relationships. An operation is defined by an individual doing something to the physical world. Reflective abstractions made by the epistemic subject are also individualized; no communication with other epistemological or psychological subjects is needed—an assumption also underlined by Piaget in the idea that the sources of mathematics and logic are "deeper" and different from the sources of language. I find similarities between the *ratio* described by rationalistic philosophy and the epistemic subject. Both have a common nature, and both—in their pure form—work in isolation. The genetic epistemology of Piaget is monological.

Let us summarize: A monological epistemological theory is characterized by its interpretation of the genesis of (mathematical) knowledge. It is described in concepts that make reference only to the individual and to some specific experiences upon which the individual can establish knowledge. Perhaps the individual must pass some stages on its way towards a more comprehensive understanding, but in the description of the various stages and of how to pass from one stage to the following, references are made only to intellectual structures of the individual or to relationships between the individual and some part of the physical world. Of course, a monological epistemology does not claim that knowledge development in fact takes place in private or that knowledge development ought to take place in private. But it explains knowledge development by means of concepts that, from a logical point of view, are oriented towards the individual.

The main questions now are, of course, how to outline an alternative epistemological perspective; how to understand the political dimension of mathematical education; and how to reveal the political dimensions of mathematical education behind a monological epistemology?

MONOLOGISM IN MATHEMATICAL EDUCATION

The monological character of the epistemology used for interpretation and development in mathematical education has drawn the attention of teachers and curriculum developers towards some aspects and simultaneously created blindness with regard to other aspects of what is going on in the classroom.

In primary school teaching monologism puts the intellectual developments of the child in focus. The child must be placed in the most stimulating environments and, with reference to the specific analysis of Piaget, it is possible to specify important features of these environments. It should be possible for the child to be involved in concrete operations, facilitating the development of logico-mathematical patterns of thought.

The possibility for specifying goes much further. Piaget finds that the basic structures of operations are similar to the three basic structures in mathematics, which have obtained the status of mother structures in the architecture of mathematics developed by Bourbaki. The structures of operations are interpreted as the genetic roots of the mathematical mother structures; and therefore it seems possible to develop a variety of concrete mathematical structures starting with the most elementary manipulations of blocks and bricks and finally reaching the abstract mathematical structures.

The planning of this sequence of teaching/learning material can take place outside the classroom. The goal is fixed by mathematics itself in the formal structures that, from a logical point of view, are basic to all mathematics, and the path is outlined by the stages that the child has to pass on its way toward logical maturity.

And in fact a great deal of teaching/learning material was produced during the 1960s and in the beginning of the 1970s with that epistemological paradigm in mind. What the child has to do is outlined in great detail. The basic teaching/learning relationship is that between the textbook and the child, or between the textbook and the teacher and the child. Less attention is paid to the epistemological potential in the relationship between the children.

The communication among children and between children and their teacher has a role to play. It is not claimed that this is a minor role or that it is unimportant for a good teacher to keep in close contact with the children. The point is that the monological epistemology does not underline interpersonal relationships as essential to the genesis of knowledge. Communication is reduced to a pedagogical and a methodological concept.

In normal secondary-school teaching monologism legalizes mathematical education in terms of information: the teacher informs the students about mathematical facts, and the good teacher accommodates each explanation to the specific knowledge previously acquired by students. Monologism establishes authorities in the classroom—the teacher and the textbook.

This reveals another aspect of traditional epistemology connected to education. This aspect accompanies monologism and relates to the structure of knowledge. By looking into mathematical classrooms, illuminated by a monological epistemology, it becomes obvious that monologism is closely connected to the assumption that a coherent source of knowledge exists. A unique basis for knowledge makes consistency of knowledge possible, and we reach the thesis of homogeneity of knowledge.

HOMOGENEITY OF KNOWLEDGE

The thesis of *homogeneity of knowledge* states that it is possible to integrate knowledge into one unified system. This thesis is one of the distinct features of logical positivism, and is maintained, for example, by Rudolf Carnap. In the paper "Die alte und die neue Logik" from 1930, he states that an analysis of the concepts in science will show that, no matter to which subject the concepts belong in the traditional classification of physics, biology, chemistry, psychology, and so forth, they will find their place in a unified system containing all scientific knowledge. This idea finds support in rationalism and in empiricism, which both identify a uniform source of all types of knowledge. Further, it is supported by the classical interpretation of knowledge. This interpretation is found in the work of Plato, in the whole tradition of epistemology, and, for instance, also in the epistemology developed by Bertrand Russell.

According to this interpretation, three conditions must be fulfilled if a person is going to have knowledge. The person A knows p if and only if: (1) A believes that p is true, (2) A has sufficient reasons to justify p, and (3) p is true.

Therefore, if a person knows p, p must be true. This is why it makes sense to maintain that all knowledge could become integrated into one common system; true sentences cannot contradict each other. This basic assumption is, for instance, developed by Ludwig Wittgenstein in *Tractatus logico-philosophicus* in the thesis that an arbitrary proposition is a truth-function of elementary propositions not capable of contradicting each other.

The thesis of homogeneity does not imply a monologism; neither does monologism imply a thesis of homogeneity. Monologism deals with how to obtain knowledge, and the thesis of homogeneity deals with the structure of knowledge. However, there is an affinity between the two ideas. In rationalistic philosophy we meet the assumption that rational capacity is of a uniform logical nature combined with the assumption that by rational means we will be able to enter the whole area of knowledge. We find a similar idea in Piagetian thought. The "epistemic subject" is common to everybody, which is the reason why the individual is able to obtain mathematical knowledge, if it is placed in an appropriate environment (one in which objects can be operated with). Interaction does not have any essential role to play.

In the normal mathematical classroom harmony exists. Mathematics is conceived as a single homogeneous system of knowledge. And, naturally, the aim of teaching must be to inform the students as well as possible. The teacher has to introduce the students to the subject, and due to monologism the focus is on the relationship between the individual student and that parcel of the curriculum that is presented by the textbook and explained by the teacher. In that way homogeneity of knowledge introduces authority into the classroom.

KNOWLEDGE CONFLICTS

However, I find the thesis of homogeneity of knowledge problematic. I find it necessary to take a new approach to the definition of knowledge. My inspiration comes from the philosophy of language developed by Wittgenstein in *Philosophische Untersuchungen*, and from J. L. Austins theory of speech acts. I do not combine knowledge with truth in any absolute way, but with a "willingness" to argue in favor of a specific position. Further, I combine knowledge with promise. If I maintain that I know p, I also assure that I am in possession of a sufficient background to maintain p. I must be sure that p is true. "To be sure," however, is not a metaphysical concept. It means that p is true in all the situations I am able to imagine.

Let us state these conditions in a more systematic way. A person A knows p if and only if:

1.   A guarantees that he/she understands p, that is, that he/she is able to interpret the concepts used in the formulation of p
2.   A guarantees that he/she has tried to imagine a situation in which p is false but found the situation impossible to realize
3.   A guarantees that the set of situations that A is able to imagine is broad enough to cover all relevant possibilities
4.   A guarantees that others could act as if p is true
5.   A has a willingness to be placed under a responsibility for discussion, that is, he/she intends to explain why hypothetical situations in which p is false never can appear.

In this interpretation we are in accord with the natural language use of *knowledge,* and also with modern logical analysis inspired by intentional logic, developed from modal logic, and using a "possible world" semantics (see for example, Hintikka, 1975).

This definition of knowledge has similarities with "promise." A person who declares: "I promise *p,*" is not only making a statement, but also acting. This person is promising and, in doing so, carrying out a speech act.

This speech-act dimension is also found in the concept of knowledge, although ignored in the whole tradition of epistemology focusing on the classical three-step definition. This definition presupposes that a knowledge statement is totally descriptive. It informs us about a person's belief, about a logical structure creating sufficient reason for the person to maintain a proposition, and about a state of affairs contained in the proposition in question. The speech-act interpretation draws attention to the fact that a knowledge statement includes an act, and to why an epistemological analysis of knowledge must include intentional concepts (like 'assure', 'promise', and so on). That is why I call my knowledge interpretation *intentional,* as opposed to the classical *descriptive* interpretation of knowledge.

The expression "A knows p" is no more specific than "A promises p"; this is obvious from the fact that in our interpretation of knowledge we have to use the concept 'guarantee'. This concept does not have any well-defined use, and neither does 'knowledge'. We have to take the concept 'knowledge' away from metaphysical dreams about eternal truths, sufficient reasons, and so on.

The intentional interpretation implies that we have to face *knowledge conflicts*. They belong to the very heart of knowledge development. By knowledge conflicts I do not mean belief conflicts between two people who have different opinions. Such conflicts are unproblematic from an epistemological point of view. That people may have different beliefs is obvious. In the descriptive definition, knowledge conflicts are impossible. What I want to underline is that the intentional definition makes the existence of knowledge conflicts possible—without degenerating the concept of 'knowledge' into a concept of 'belief'.

The logical structure of a knowledge conflict is: (1) A knows p, (2) B knows q, and (3) p and q make up a contradiction. I am denying the existence of any divine authority. It is impossible to identify any final method for solving a knowledge conflict. Instead, a knowledge conflict makes interaction necessary.

The possibility exists that knowledge conflicts may exist somewhere, although not in some specific areas of knowledge—for instance, mathematics. Perhaps mathematics constitutes an exceptional and large homogeneous body of knowledge? This implies that it is unproblematic in mathematic education to let monologism accompany harmony.

The thesis of harmony (dressed in monologism) creates the blind spots that make it impossible to discuss the political dimensions of mathematical education from an epistemological point of view. The thesis of harmony creates the idea of neutrality and objectivity of mathematical calculations and solutions to practical problems. When the thesis of harmony enters the classroom, disagreement must be a sign of some misunderstandings.

When a knowledge system is beyond doubt, when no possibility for knowledge conflicts exists, it becomes sacramental. The sacred position of mathematical knowledge is the basic epistemological cause of the invisibility of its political dimensions.

So, the main question must be: Is it possible to identify knowledge conflicts in mathematics? And if the answer is yes: What is the nature of knowledge progression from a situation of knowledge conflicts?

TYPES OF KNOWLEDGE

The elimination of knowledge conflicts in mathematical education constitutes the starting point for some of the ideological functions of mathematics. Here we find the origin of the objectivism of mathematics. (For a discussion of this concept, see Bishop, 1988.)

One of the main problems related to mathematical modelling I have summarized elsewhere as the concealment of preunderstanding, to be understood as the phenomenon, accompanying a mathematization, of disguising the complexity of the construction of the conceptual system that constitutes the very foundation of the model itself (Skovsmose, 1990b). Another problem concerns the confusion about the goals of modelling. A model could be used to make predictions, or to make descriptions, or it could be used in a context of discovery, and so forth. And the model has to be discussed in different ways, depending on the intended application of the model (see Skovsmose 1990b).

The nature of mathematical language makes it tempting to accept a "picture"-theory of language as developed by Wittgenstein in the *Tractatus*; by doing this, an object to be pictured by the mathematical model is invented. That is the phenomenon of "objectivization" connected to mathematics, leading immediately to the idea that mathematics constitutes a suitable language for stating value neutral descriptive facts. To attack this point of view we have to develop an idea of *reflective knowledge*. (For a discussion of the concept of 'reflective knowledge', see Skovsmose 1989b, 1990a, 1990b; also Boos-Bavnbek and Pate, 1989; Hanna, 1989.)

It is possible to make a quite fundamental distinction between technological knowledge and reflective knowledge. The concept of 'technology' is normally interpreted as referring both to technological "machinery" (for example, steam engine, computers and so on), and to technological "know-how" concerning how to use and how to construct the technological hardware. Technological know-how I call "technological knowledge." My fundamental thesis is that technological knowledge normally does not include knowledge about how to discuss the results and functions of technological constructions. The knowledge that is necessary in the construction of computer systems is not sufficient for evaluating those systems. Knowledge of this broader type I call "reflective knowledge."

Technological knowledge and reflective knowledge constitute two different types of knowledge, but not two independent types. It is important to master some technological knowledge to develop reflective knowledge. However, it is not necessary to master all aspects of technological knowledge. If that were the case, democracy in a highly technological society would become impossible; only experts would become able to control experts.

In addition, it is necessary to make a distinction between mathematical knowledge itself and technological knowledge. This is obvious in education where students will become unable to use mathematics in modelling processes if a specific concern is not shown in the process of mathematization. A difference exists between mathematical competence and competence in the application of mathematics. The implication is that we have to make distinctions among (1) mathematical knowledge itself; (2) technological knowledge, including the ability to use mathematics; and (3) reflective knowledge, concerning the criticism of the results of technological constructions. From an epistemological point of view the essential question is: What is the nature of reflective knowledge?

I find it impossible to maintain the idea of the existence of reflective knowledge without accepting the possibility for knowledge conflicts. We cannot expect to identify a specific source for reflective knowledge.

KNOWLEDGE CONFLICTS IN MATHEMATICAL EDUCATION

An important task in education is to develop the foundations for reflective understanding. It is especially important in mathematical education, because mathematics constitutes an important part of a great variety of today's technological knowledge.

However, no concern has been shown about reflective knowledge in traditional mathematical education. This has to do with educational traditions, which again have been incorporated into the monological epistemological paradigm. Homogeneity is created and authority introduced by the elimination of reflective knowledge.

Reflective knowledge does not constitute a simple development of technological knowledge. A critical approach to that type of knowledge cannot be established without accepting the possibility of knowledge conflicts. Questions leading to reflective knowledge related to the application of some specific areas of mathematics could concern:

1. The applicability of the subject: Who uses it? Where is it used?
2. The assumptions behind the applications: Which theoretical assumptions and which simplifications have made the mathematical model construction possible?
3. The interest behind the application: Which knowledge constituting interests are connected to the development of the subject? Is the application in question guided by political or other interests?
4. The functions of the applications: Are the mathematical results

used in any specific way that may create unexpected consequences?

These questions only indicate the nature of reflective knowledge. It could become oriented towards the most advanced use of mathematical models in modern technology.

Also, it could become oriented toward quite elementary uses of mathematics. In this case reflective knowledge could emerge if the curriculum were interpreted in the light of the conceptions of the students. That means that reflective knowledge concerns the process of learning mathematics, and questions could be:

1. Why do we (the students) have to learn the subject in question? Will it become useful for us, or is it necessary to learn only because of examinations, and so forth?
2. Is it possible for us (the students) to describe and explain the content of the curriculum without using the concepts from the curriculum?
3. How does the subject in question fit into the 'curriculum of the students', to be interpreted as the well-elaborated (but often unconscious) program for what the students want to learn? (The concept 'curriculum of the students' is developed by Lena Lindenskov [1990].)

Both when reflective knowledge concerns advanced uses of mathematics and when it concerns learning we face knowledge conflicts. No unified framework is able to contain concepts from mathematics and at the same time concepts for analyzing the use of the mathematical concepts. Mathematics could become formalized, but it could not become the basis for evaluating that formalization.

To accept harmony of knowledge means that reflections about formalization will become eliminated; but the development of reflective knowledge is presupposed if we are going to reveal the political dimension of mathematical education.

THE DIALOGICAL NATURE OF KNOWLEDGE CONFLICTS

An important feature of mathematical knowledge presented in a standard curriculum is that knowledge conflicts cannot emerge. A disagreement will always indicate that somebody has misunderstood something. However, the characteristic of a knowledge conflict is precisely that it is a conflict between areas of knowledge (to be interpreted not in the descriptive but in the intentional way).

Conflicts have to do with the impossibility of combining and comparing in a homogeneous conceptual framework. In the classical interpretation we could imagine the existence of a single homogeneous knowledge system including all truths. However, if knowledge conflicts emerge we have instead to deal with only a patchwork. We always have to remedy our knowledge. No "ups," no "downs," exist in any structure of knowledge. No unique hierarchical structure is available.

A knowledge conflict cannot be solved by any additional information, observation, or by any simple act. The question is not only how to improve the conditions for observation. This would be an empiricist way of handling the situation. Neither is it possible to dissolve a knowledge conflict by further information. A knowledge conflict has to be handled in another way. It cannot be interpreted by a monological theory of epistemology.

If a knowledge conflict should involve a dynamical process, its dialogical nature has to be underlined. In the description of knowledge, we must use intentional concepts.

Knowledge is a dispositional concept indicating possible speech acts to be carried out in case the person A is put into specific interpersonal relationships. It is not possible to construct a definition of knowledge without the concept of dialogical relationships. The intentional concepts do not make any sense if the situation is monological.

It is essential that we develop a (new) genetic epistemology, if we are going to be able to describe what is going on in the development of knowledge, and to capture the idea of knowledge conflicts. Such conflicts indicate that a dialogical relationship has to be established. If A knows p, B knows q, and p and q make up a contradiction, any progression will depend on the interaction between A and B. That is the only way to handle a situation with two partial conceptual systems.

Knowledge conflicts indicate that some conceptual frameworks are incompatible, perhaps inconsistent. Therefore, a knowledge conflict must result in a *change of concepts*.

One possibility is that concept development could become systematized, that is, a well-specified procedure may create new concepts necessary for coping with the conflicting situation. We find attempts to elaborate such algorithms as part of the "classical program" in artificial intelligence. The aim was to create systems at the same time intelligent and mechanical, able to develop concepts useful for incorporating new information. In fact we could identify monologism, maintaining the possibility for some sort of automatic knowledge production.

The other possibility is to accept that no mechanical procedure exists for dissolving knowledge conflicts. Conceptual change could take place only in a process such as the one articulated by Lakatos. The history of proofs and refutations underlines the fact that the creation of a concept presupposes the identification of different possible states of affairs, combined with a decision as to whether the concept in question has to cover the situation or not.

Concept development is a dialogical process. The intentional interpretation of knowledge implies that if knowledge is maintained, a person A (or a group of persons) will "promise" that it is possible for others to act as if something, stated in a proposition p, is true, and A will be placed under an obligation to argue for the truth of p. The denial of p could be supported by the identification of a situation, not conceived by A and, at the same time, exemplifying that p may become false. However, the identification of such a hypothetical situation involves conceptual creativity. It has also to be evaluated if new situations have to be taken into consideration, if it is relevant to the presupposed definition of the concept used in the formulation of p.

My thesis is that we will be able to develop reflective knowledge only if we move into dialogical situations like the one outlined. That means that if we want to grasp the nature of formalization we must be involved in a dialogical process. This will be the case if we have to deal with more advanced examples of mathematics in use, as well as with a commonsense interpretation of what is going on in an elementary mathematical calculation. In both cases the object of our reflections is some fragment of knowledge. We try to develop reflective knowledge. The ideological function of monologism is to rule out the importance of dialogue, and, by doing this, the possibility for knowledge conflicts being the starting point of reflections will become eliminated.

The introduction of reflective knowledge as an essential part of educational practice means the introduction of knowledge conflicts into the classroom. Therefore monologism has to be replaced by a dialogical interpretation of knowledge development; this will break down the ideology of the neutrality of the mathematical curriculum.

CONCLUSIONS

Monologism draws our attention to the learner as an individual person, making a totally individualized education possible—at least from a logical point of view. Monologism supports and is supported by the idea that knowledge could become incorporated into one homogeneous structure, which is summarized in the descriptive interpretation of knowledge.

The intentional interpretation makes knowledge conflicts possible. If we have to proceed from knowledge conflicts, concept development is necessary, but this cannot become a mechanical enterprise. Dialogue will become the logic of concept development.

If these possibilities are excluded from mathematical education, it is reduced to a guided trip for students into an ordered world of mathematical entities, without any possibility of their developing a critical sense of what they have to learn. The whole subject will emerge as an objective and value free structure. To overcome this picture it is necessary—from an epistemological point of view—to identify the political dimension of mathematical education.

If that is done it makes sense to evaluate the functions of the use of formalized theories in society, and it makes sense to confront the content of mathematical education with the ideas and expectations of the students. And it makes sense to interpret mathematical education as a part of a dominating culture, which could ruin other cultures but which—at the same time—could be criticized and changed. However, I am not saying anything about what the specific investigations may result in. I have only tried to point out that monologism and the thesis of harmony create blind spots behind which we find the political dimension of mathematical education. Next, I have pointed out that a useful epistemological step for moving into an analysis of that dimension is the realization of the dialogical nature of reflective knowledge.

NOTE

This chapter is developed as part of the project "Mathematical Education in an Epistemological Perspective" (MEEP), which is supported by the Danish Research Council for the Humanities.

REFERENCES

D'Ambrosio, U.

   1981 "Uniting Reality and Action: A Holistic Approach to Mathematics Education." Pp. 33-42 in L. A. Steen and D. J. Albers (eds) *Teaching Teachers, Teaching Students*. Boston, Basel, Stuttgart: Birkhauser.

   1985 "Mathematical Education in a Cultural Setting." *International Journal of Mathematical Education in Science and Technology* 16:469-77.

Austin, J. L.

   1962 *How to Do Things with Words*. Oxford: Oxford University Press.

   1970 *Philosophical Papers*. 2nd ed. Oxford: Oxford University Press.

Beth, E. W., and J. Piaget

   1966 *Mathematics, Epistemology and Psychology*. Dordrecht: D. Reidel.

Bishop, A.

   1988 *Mathematical Enculturation*. Dordrecht, Boston, and London: Kluwer.

Bloor, D.

   1988 *Knowledge and Social Imagery*. London: Routledge and Kegan Paul.

Blum, W., J. S. Berry, R. Bieler, I. D. Huntley, G. Kaiser-Messmer, and L. Profke (eds.).

   1989 *Applications and Modelling in Learning and Teaching Mathematics*. Chichester: Ellis Harwood.

Boos-Bavnbek, B., and G. Pate

   1989 "Information Technology and Mathematical Modelling, the Software Crisis, Risk and Educational Consequences." *Zentralblatt für Didaktik der Mathematik* 89, 5:167-75.

Carnap, R.

   1930 "Die alte und die neue Logik" *Erkenntnis* 1. Translated into English by Isaac Levi; pp. 133-46 in A. J. Ayer (ed.), 1959, *Logical Positivism*. New York: Free Press.

Fischer, R.

   1989 "Social Change and Mathematics." Pp. 12-21 in Blum et al. (1989).

Fischer, R., G. Malle, with H. Burger

   1985 *Mensche und Mathematik: Eine Einfuhrung in didaktisches Denken und Handeln*. Mannheim: Bibliographisches Institut.

Hanna, G.

   1989 "Teaching the Appropriate Use of a Mathematical Model." Pp. 312-16 in Blum et al. (1989).

Hintikka, J.
1975 *The Intentions of Intentionality and Other New Models for Modalities.* Dordrecht: D. Reidel.

Keitel, C.
1989 "Mathematics Education and Technology." *For the Learning of Mathematics* 9:7-13.

Lakatos, I.
1976 *Proofs and Refutations.* Cambridge: Cambridge University Press.

Lindenskov, L.
1990 "The Pupil's Curriculum: Rationale of Learning and Learning Stories." pp. 165-172 In Noss et al. (1990).

Mellin-Olsen, Stieg
1981 "Instrumentalism as an educational concept." *Educational Studies in Mathematics,* 12:351-67.
1987 *The Politics of Mathematics Education.* Dordrecht, Boston: D. Reidel.

Mill, J. S.
1970 *A System of Logic.* London: Longman.

Noss, R. et al., eds.
1990 *Political Dimensions of Mathematical Education: Action and Critique.* Proceedings of the First International Conference, Institute of Education, University of London.

Piaget, J.
1970 *Genetic Epistemology.* New York: Columbia University Press.

Russell, B.
1948 *Human Knowledge.* London: George Allen and Unwin.

Skovsmose, O.
1985 "Mathematical Education Versus Critical Education." *Educational Studies in Mathematics* 16:337-54.
1988a "Mathematics as Part of Technology." *Educational Studies in Mathematics* 19:23-41.
1988b *Reflective Knowledge and Mathematical Modelling.* R 88-13. Department of Mathematics and Computer Science, Aalborg University Centre.
1988c *Democratization and Mathematical Education.* R 88-33. Department of Mathematics and Computer Science, Aalborg University Centre.
1989a "Towards a Philosophy of an Applied Oriented Mathematical Education." Pp. 110-114 in Blum et al. (1989a).
1989b "Models and Reflective Knowledge." *Zentralblatt für Didaktik der Mathematik* 89, 1:3-8.
1989c *Perspective on Curriculum Development in Mathematical Education.* R. 89-41. Department of Mathematics and Computer Science, Aalborg University Centre.

1990a "Mathematical Education and Democracy." *Educational Studies in Mathematics* 21:109-28.
1990b *Reflective Knowledge—Its Relation to the Mathematical Modelling Process.* R 90-14. Department of Mathematics and Computer Science, Aalborg University Centre.

Wittgenstein, L.
1922 *Tractatus logico-philosophicus.* London: Routledge & Kegan Paul.
1953 *Philosophische Untersuchungen.* Oxford: Blackwell.

PHILIP J. DAVIS

10

Applied Mathematics as
Social Contract

THIS MATHEMATIZED WORLD

As compared to the medieval world or the world of antiquity, today's world is characterized as being scientific, technological, rational, and mathematized. By "rational" I mean that by an application of reason or of the formalized versions of reason found in mathematics, one attempts to understand the world and control the world. By "mathematized," I mean the employment of mathematical ideas or constructs, either in their theoretical form or in computer manifestations, to organize, to describe, to regulate, and to foster our human activities. By adding the suffix "-ized", I want to emphasize that it is humans who, consciously or unconsciously, are putting the mathematizations into place and who are affected by them. It is of vital importance to give some account of mathematics as a human institution, to arrive at an understanding of its operation and at a philosophy consonant with our experience with it and, on this basis, to make recommendations for future mathematical education.

The pace of mathematization of the world has been accelerating. It makes an interesting exercise for young students to count how many numbers are found on the front page of the daily paper. The mere number of numbers is surprising, as well as the diversity and depth of the mathematics that underlies the numbers; and if one turns to the financial pages or the sports pages, one sees there natural language overwhelmed by digits and statistics. Computerization represents the effective means for the realization of current mathematizations as well as an independent driving force toward the installation of an increasing number of mathematizations.

PHILOSOPHIES OF MATHEMATICS

Take any statement of mathematics such as "two plus two equals four" or any more advanced statement. The common view is that such a statement is perfect in its precision and in its truth, is absolute in its objectivity, is universally interpretable, is eternally valid, and expresses something that must be true in this world and in all possible worlds. What is mathematical is certain. This view, as it relates, for example, to the history of art and the utilization of mathematical perspective has been expressed by Kenneth Clark: The Florentines demanded more than an empirical or intuitive rendering of space. "They demanded that art should be concerned with *certezza*, not with *opinioni*." (Clark, 1961:20). *Certezza* can be established by mathematics.

The view that mathematics represents a *timeless* ideal of absolute truth and objectivity and is even of nearly divine origin is often called "Platonism." It conflicts with the obvious facts that we humans have invented or discovered mathematics, and that we have installed mathematics in a variety of places both in the arrangements of our daily lives and in our attempts to understand the physical world. In most cases, we can point to the individuals who did the inventing or made the discovery or the installation, citing names and dates. Platonism conflicts with the fact that mathematical applications are often *conventional* in the sense that mathematizations other than the ones installed are quite feasible (for example, the decimal system). The applications are often *gratuitous*, in the sense that humans can and have lived out their lives without them (for instance, insurance or gambling schemes). They are *provisional* in the sense that alternative schemes are often installed that are claimed to do a better job. (Examples range all the way from tax legislation to Newtonian mechanics.) Opposed to the Platonic view is the view that a mathematical experience combines the external world with our interpretation of it, via the particular structure of our brains and senses, and through our interaction with one another as communicating, reasoning beings organized into social groups.

The perception of mathematics as quasi-divine prevents us from seeing that we are surrounded by mathematics because we have extracted it out of unintellectualized space, quantity, pattern, arrangement, sequential order, and change and that, as a consequence, mathematics has become a major modality by which we express our ideas about these matters. The conflicting views, as to whether mathematics exists independently of humans or whether it is a human phenomenon, and the emphasis that tradition has placed on the former view lead us to shy away from studying the process of mathematization, to shy away from asking embarrassing questions about this process: how do we install the mathematizations? why do we install them? what are they doing for us or to us? do we need them? do we want them? on what basis do we justify them? But the discussion of such questions is becoming increasingly important as the mathematical vision transforms our world, often in unforeseen ways, as it both sustains and binds us in its steady and unconscious operation. Mathematics creates a reality that characterizes our age.

The traditional philosophies of mathematics, Platonism, logicism, formalism, intuitionism, in any of their varieties, assert that mathematics expresses precise, eternal relationships between atemporal mental objects. These philosophies are what Thomas Tymoczko has called "private" theories. In a private theory, there is one ideal mathematician at work, isolated from the rest of humanity and from the world, who creates or discovers mathematics by his own logico-intuitive processes. As Tymoczko points out, private theories of the philosophy of mathematics provide no account either for mathematical research as it is actually carried out, for the applications of mathematics as they actually come about, or for the teaching process as it actually unfolds. When teaching goes on under the banner of conventional philosophies of mathematics, it often becomes a formalist approach to mathematical education: do this, do that, write this here and not there, punch this button, call in that program, apply this definition and that theorem. It stresses operations. It does not balance operations with an understanding of the nature or the consequences of the operations. It stresses syntactics at the expense of semantics, form at the expense of meaning. A fine place to read about this is in *L'age du capitaine* by Stella Baruk, a mathematics supervisor in a French school system. Baruk writes: "From Pythagoras in antiquity to Bourbaki in our own days, there has been maintained a tradition of instruction—religion which sacrifices full understanding to the recitation of formal and ritual catechisms, which create docility and which simulate sense. All this has gone on while the High Priests of the subject laugh in their corners." (Baruk, 1985, my translation).

How many university lecturers, discoursing on numbers, say, allow themselves to discuss where they think numbers come from, what one's intuition is about them, how number concepts have changed, what applications they have elicited, what pressure has been exerted by applications, how we are to interpret the consequences of these applications, what the poetry of numbers is or their drama or their mysticism, why there can be no complete or final understanding of them? How many lecturers would take time to discuss the question put by Bertrand Russell in a relaxed moment: "What is the Pythagorean power by which number holds sway above the flux?"

Opposed to "private" theories, there are "public" theories of the philosophies of mathematics in which the teaching process is of central importance. Several writers in the past half century have been constructing public theories, and I should like to add a few bricks to this growing edifice and to point out its relevance for the future of mathematical education.

APPLIED MATHEMATICS AS SOCIAL CONTRACT

I shall emphasize the applications of mathematics to the social or humanistic areas, though one can make a case for applications to scientific areas and indeed to pure mathematics itself (see, e.g., Spalt, 1986).

Today's world is full of mathematizations that were not here last year or ten years ago. There are other mathematizations that have been discarded (for example, Ptolemaic astronomy, numerological interpretations of the cosmos, last year's tax laws). How do these mathematizations come about? How are they implemented, why are they accepted? Some are so new, for example, credit cards, that we can actually document their installation. Some are so ancient, for example, numbers themselves, that the historical scenarios that have been written are largely speculative. Are mathematizations put in place by divine fiat or revelation? By a convention of elders? By the insights of a gifted few? By an evolutionary process? By the forces of the market place or of biology? And once they are in place what keeps them there? Law? Compulsion? Inertia? Darwinian advantage? The development of bureaucracy whose sole function it is to maintain the mathematization? The development of business whose function it is to create and sell the mathematization? Well, all of the above, at times, and more. But for all the lavish attention that our historians of mathematics have paid to the evolution of ideas within mathematics itself, only token attention has been paid by scholars and teachers to the interrelationship between mathematics and society. A description of mathematics as a human institution would be complex, indeed, and not easily epitomized by a catch phrase or two.

The employment of mathematics in a social context is the imposition of a certain order, a certain type of organization. Government, as well, is a certain type of organization and order. Philosophers of the seventeenth and eighteenth centuries (Hobbes, Locke, Rousseau, Thomas Paine, and so on) put forward an idea, known as 'social contract,' to explain the origin of government. Social contract is an act by which an agreed-upon form of social organization is established. (Here I follow an article by Michael Levin [1973].) Prior to the contract there was supposedly a "state of nature." This was far from ideal. The object of the contract, as Rousseau put it, was "to find a form of association which will defend the person and goods of each member with the collective force of all, and under which each individual, while counting himself with the others, obeys no one but himself, and remains as free as before" (Levin, 1973:259). In this way, one may improve on a life that, as Hobbes put it in a famous sentence, was "solitary, poor, nasty, brutish, and short." The contract itself, whether oral or written, was almost thought of as having been entered into at a definite time and place. Old Testament history, particularly God's covenants with Noah, Abraham, Moses, the Children of Israel, was clearly in the minds of contract theorists. In the United States, political thinking has often been in terms of contracts, as in the Mayflower Compact, the constitutions of the United States and of the individual states, the establishment of the United Nations in San Francisco in 1945, and periodic proposals for constitutional amendments and reform.

It was generally assumed by the contract theorists that "Human society and government are the work of man constructed according to human will even if sometimes operating under divine guidance." That "man is a free agent, rather than a being totally determined by external forces," and that society and government are based on mutual agreement rather than on force (Levin, 1973).

The acceptability of social contracts as a historical explanation hardly lasted till the nineteenth century, even if political contracts continued to be entered into as instances of democratic polity. It is an instructive exercise, I believe, in order to get a grasp on the relationship between society and mathematics, to take the outline of social contract just given and replace the words *government and society* by the word *mathematization*. Though it is naive to think that most mathematizations came about by formal contracts, the "contract" metaphor is a useful phrase to use in designating the interplay between people and their mathematics and to make the point that mathematizations are the work of man, constructed according to human will, even if operating under a guidance that may be termed divine or logical or experimental, according to one's philosophic predilection.

A number of authors, some writing about theology and others about political or economic processes, have pointed out that contracts are continuously entered into, broken, and reestablished. I believe the same is the case for mathematizations. Consider, for example, insurance. This is one of the great mathematizations currently in place, and I personally, without adequate coverage, would consider myself naked to the world. Yet I am free to throw away my insurance policies. Consider the riders that insurance companies send me, unilaterally abridging their previous agreements. Consider also that in a litigious age, with a populace abetted by eager lawyers and unthinking juries, what appears as the "natural" stability of the averages upon which the possibility of insurance is based, emerges, on deeper analysis, to incorporate the willingness of the community to adjust its affairs in such a way that the averages are maintained. The possibilities of insurance can be destroyed by our own actions.

Another example that displays the relationship between mathematics, experience, and law is the highway speed limit in the United States. Before the gas shortage in 1974, the limit was sixty-five miles an hour. In 1974, the speed limits were reduced to fifty-five miles an hour in order to conserve gasoline. As a side effect, it was found that the number of highway accidents was reduced significantly. The gas shortage ended in 1987, and since then there has been pressure to raise the legal speed limit. Society must decide what price it is willing to pay for what some see as the convenience or the thrill of higher speeds. Here is mathematical contract at work.

The process of contract maintenance, renewal of reaffirmation, in all its complexities, is open to study and description. This is a proper part of applied mathematics, and I shall argue that it should be a proper part of mathematical education.

WHERE IS KNOWLEDGE LODGED?

There is an epistemological approach to the interplay between mathematics and society, and that is to look at the way society answers the question that heads this section. According to how we answer this question, we will mathematize differently and we will teach differently.

Where, then, is knowledge lodged? (Here I follow an article by Kenneth A. Bruffee [1982]). In the pre-Cartesian age, knowledge was often thought to be lodged in the mind of God. Those who imparted knowledge authoritatively derived their authority from their closeness to the mind of God, and evidence of this closeness was often taken to be the personal godliness of the authority.

In the post-Cartesian age knowledge was thought to be lodged in some loci that are above and beyond ourselves, such as sound reasoning or creative genius or in the "object objectively known."

A more recent view, connected perhaps with the names of Kuhn and Lakatos, is that knowledge is socially justified belief. In this view, knowledge is not located in the written word or in symbols of whatever kind. It is located in the community of practitioners. We do not create this knowledge as individuals, but we do it as part of a belief community. Ordinary individuals gain knowledge by making contact with the community experts. The teacher is a representative of the belief community.

In my view, knowledge as socially justified belief provides a fair description of how mathematical knowledge is legitimized, but we must keep clearly in mind that perceptions of what "is," theory formation, validation, and utilization are all part of a dynamic and iterative process. Knowledge once thought to be absolute, indubitable, is now seen as provisional or even probabilistic. Science is seen as a search for error as much as it is a search for truth. Eternally valid knowledge may remain an ideal that we hold in our minds as a spur to inquiry. This view fits with the idea of applied mathematics as social contract, with the contractual arrangements being concluded, broken, and renegotiated in endless succession.

Another view of the locus of knowledge, a view not yet elevated to a philosophy, is that knowledge is located in the computer. One speaks of such things as "artificial intelligence" and "expert systems," and more than one theoretical physicist has opined that all the essentials are now known (despite the fact that the same was asserted a hundred years ago and two hundred years ago) and that the computer can fill in the details and derive the consequences tor the future.

Advocates of this view have asserted that while education is now teacher oriented, in the full bloom of the computer age, education will be knowledge oriented. These two contemporary views are not necessarily antithetical, provided we accommodate the computer into the community of experts, clarify whether "belief" can reside in a computer, and decide whether humankind exists for the sake of the computers or vice versa.

MATHEMATICAL EDUCATION AT A HIGHER METALEVEL

A mathematized and computerized world brings with it many benefits and many dangers. It opens many avenues and closes many others. I do not want to elaborate this point, as I and my coauthor Reuben Hersh have done so in our book *Descartes' Dream*, as have numerous other authors.

The benefits and dangers both derive from the fact that the mathematical/computational way of thinking is different from other ways. Philosopher and historian Isaiah Berlin called attention to this divergence when he wrote "A person who lacks common intelligence can be a physicist of genius, but not even a mediocre historian" (Berlin, 1980). To allow the mathematical way to gain ascendancy over other modes is to create an imbalance in human life.

The benefits and dangers derive also from the fact that mathematics is a kind of language, and this language creates a milieu for thought that is hard to escape. It both sustains us and confines us. As George Steiner has written of natural language: "[The] oppressive birthright is the language, the conventions of identification and perception. It is the established but customarily subconscious unargued constraints of awareness that enslave" (1986:106). One can assert as much for mathematics as a language. The subconscious modalities of mathematics and of its applications must be made clear, must be taught, watched, argued. Since we are all consumers of mathematics, and since we are both beneficiaries as well as victims, all mathematizations ought to be opened up in the public forums where ideas are debated. These debates ought to begin in the secondary school.

Discussions of changes in mathematics curricula generally center around (1) the specific mathematical topics to be taught, for example, whether to teach the square root algorithm, or continued fractions or projective geometry or Boolean algebra, and, if so, in what grade they should be taught, and (2) the instructional approaches to the specific topics, for example, should they be taught with proofs or without? from the concrete to the abstract or vice versa? what emphasis should be placed on formal manipulations and what on intuitive understanding? with computers or without? with open ended problems or with "plug and chug" drilling?

Because of widespread, almost universal computerization, with handheld computers that carry out formal manipulations and computations of lower and higher mathematics rapidly and routinely, because also of the growing number of mathematizations, I should like to argue that mathematics instruction should, over the next generation, be *radically* changed. It should be moved up from subject-oriented instruction to instruction in what the mathematical structures and processes mean in their own terms and what they mean when they form a basis on which civilization conducts its affairs. The emphasis in mathematics instruction ought to be moved from the syntactic-logico component to the semantic component. To use programming jargon, it ought to be "popped up" a metalevel. If, as some computer scientists believe, instruction is to move from being teacher oriented to being knowledge oriented—and I believe this would be disastrous—the way in which the role of the teacher can be preserved is for the teacher to become an interpreter and a critic of the mathematical processes and of the way these processes interact with knowledge as a database. Instruction in mathematics must enter an altogether new and revolutionary phase.

THE INTERPRETIVE COMPONENT OF
MATHEMATICAL EDUCATION

Let me begin by asking the question: to what end do we teach mathematics? Over the millennia, answers have been given and they have differed. Some of them have been: we teach it for its own sake, because it is beautiful; because it reveals the divine; because it helps us think logically; because it is the language of science and helps us to understand and reveal the world; because it helps our students to get a job either directly, in those areas of social or physical sciences that require mathematics, or indirectly, insofar as mathematics, through testing, acts as a social filter, admitting to certain professional possibilities those who can master the material. We teach it also to reproduce ourselves by producing future research mathematicians and mathematics teachers.

Ask the inverse question: what is it that we want students to learn? We may answer this by citing specific course contents. For example, we may say that we now want to emphasize discrete mathematics as opposed to continuous mathematics. Or that we want to develop a course in nonstandard analysis on tape so that joggers may learn about hyperreal numbers even as they run. Or, we may decide for ourselves what the characteristic, constitutive ingredients of mathematical thought are: space, quantity, deductive structures, algorithms, abstraction, generalization, and so forth, and simply assure that the student is fed these basic ingredients, like vitamins. All of these questions and answers have some validity, and tradeoffs must occur in laying out a curriculum.

Within an overcrowded mathematical arena with many new ideas competing for inclusion in a curriculum, I am asking for a substantial elevation in the awareness of the applications of mathematics that affect society and of the consequences of these applications. If formal computations and manipulations can be learned rapidly and performed routinely by computer, what purpose would be served by tedious drilling either by hand or by computer? On-the-job training is certainly called for, whether at the supermarket checkout counter or on the blackboards of a high-tech development company. If mathematics is a language, it is time to put an end to overconcentration on its grammar and to study the "literature" that mathematics has created and to interpret that literature. If mathematics is a logico-mechanism of a sort, then just as only a very few of us learn how to construct an automobile carburetor, but many of us take instruction in driving, so we must teach how to "drive" mathematically and *to interpret what it means when we have been driven mathematically in a certain manner.*

What does it mean when we are asked to create insurance pools that are free of gender bias? What are the consequences when people are admitted or excluded from a program on the basis of numerical criteria? How does one assess a statement that procedure A is usually effective in dealing with medical condition B? What does it mean when a mathematical criterion is employed to judge the quality of prose or the comprehensibility of a poem or to create music in a programmatic way? What are the consequences of a computer program whose output is automatic military retaliation? The list of questions that need discussion is endless. Each mathematization-computerization requires explanation and interpretation and assessment. None of these things are now discussed in mathematics courses in the concrete form that confronts the public. If a teacher were inclined to do so, the reaction from colleagues would probably be: Well, that is not mathematics. That is applied mathematics or that is psychology or economics or social-anthropology or law or whatever. My answer would be: I am trying, little by little, to bring in discussions of this sort into my teaching. It is difficult but important.

If the claim were made, with justice, that these matters cannot be discussed intelligently without deep knowledge both of mathematics and of the particular area of the real world, then I would agree and point out that this claim forces into the open the conflict between democracy and "expertocracy" (see, e.g., Prewitt, 1983). This conflict has received considerable attention in areas such as medicine, defence, and technological pollution, but has hardly been discussed at the level of an underlying mathematical language. The tension between the two claims, that of democracy and of expertocracy, could be made more socially productive by an education that enables a wide public to arrive at deeper assessments, moving from daily experience toward the details of the particular mathematizations. While we must keep in mind certain basic mathematical material, we must also learn to develop mathematical "street smarts" that enable us to form judgements in the absence of technical expertise (cf. Prewitt, 1983).

A philosophy of mathematics which is "public" and not "private," lends support to introducing this kind of material into the curriculum. The discussion of such curriculum changes will be assisted by the perception of the mathematical enterprise as a human experience with contractual elements, and by the realization that every civilized person practices and utilizes mathematics at some level and thereby enters a certain knowledge and belief community.

Again, following Kenneth Bruffee's article (with additions and modifications), I would like to suggest a few lines of inquiry.

1. Identify and describe the mathematical beliefs, constructions, practices that are now in place. Where and how is mathematics employed in real life?
2. Describe the mathematical beliefs, constructs, and practices that have been justified by the community. What are justifiable and unjustifiable? What are the modes of justification?
3. Describe the social dimensions of mathematical practice. What constitutes a knowledge community? What does the community of mathematicians think are the best examples that the past has to offer?

As part both of the first and second points one should add: describe the nature of the various methods of prediction and the bases upon which prediction can become prescription (that is, policy).

This type of inquiry is rarely carried out for mathematics. For example, the concrete question of where such and such a piece of university mathematics is used in practical life and how widespread its use is, is seldom answered. Many textbook claims are made in text-books, but show me the real bottom line. It is important to know. How, in fact, would you define the bottom line?

The technical term for inquiries such as the above is *hermeneutics*. This word is well established in theology and in the last generation has been commonly employed in literary criticism. It means the principles or the lines along which explanation and interpretation are carried out. It is time that this word be given a mathematical context. Instruction in mathematics must enter a hermeneutic phase. This is the price that must be paid for the sudden, massive, and revolutionary intrusion of mathematizations—computerizations into our daily lives.

CONCLUSION

Mathematics is a social practice. This practice must be made the object of description and interpretation. It is ill-advised to allow the practice to proceed blindly by "mindless market forces" or as the result of the private decisions of a cadre of experts. Mathematical education must find a proper vocabulary of description and interpretation so that we can live in a mathematized world and to contribute to this world intelligently.

NOTE

I wish to thank Professor Reuben Hersh for numerous suggestions.

REFERENCES

Baruk, S.
1985 *L'âge du capitaine*. Paris: Editions du Seuil.

Berlin, I.
1980 *Against the Current: Essays in the History of Ideas*. New York: Viking.

Bruffee, K. A.
1982 "Liberal Education and the Social Justification of Belief." *Liberal Education* 68:95-114.

Clark, Kenneth
1961 *Landscape into Art*. Boston: Beacon Press.

Davis, P. J., and H. Reuben
    1981 *The Mathematical Experience.* Cambridge: Birkhauser Boston.
    1986 *Descartes' Dream: The World According to Mathematics.* San Diego: Harcourt Brace Jovanovich

Galbraith, J. K.
    1967 *The New Industrial State.* Boston: Houghton Mifflin.

Kuhn, T. S.
    1970 *The Structure of Scientific Revolutions.* 2nd ed. Chicago: University of Chicago Press.

Lakatos, I.
    1976 *Proofs and refutations.* Cambridge: Cambridge University Press.

Levin, M.
    1973 *Dictionary of the History of Ideas,* vol. 4, pp. 251-63. New York: Scribner's.

Morgenthau H.
    1972 *Science: Servant or Master:* New York: Norton.

Prewitt, K.
    1983 "Scientific illiteracy and democratic theory." *Daedalus* 112, 2 (Spring): 49–64.

Roszak, T.
    1973 *Where the Wasteland Ends: Politics and Transcendence in Post Industrial Society.* New York: Doubleday, Anchor.

Spalt, D.
    1986 "Das Unwahre des Resultatismus." Seminar für Math. und ihre Didaktik., Universität zu Köln, December.

Stanley, M.
    1978 *The Technological Conscience: Survival and Dignity in an Age of Expertise.* New York: Free Press.

Steiner, G.
    1975 *After Babel.* New York: Oxford University Press.
    1986 Review of Michel Foucault's *On Conceptual and Semantic Coercion.* *New Yorker,* March 17:105-109.

Tymoczko, T.
    1986 "Making room for mathematicians in the philosophy of mathematics.": *Mathematical Intelligencer* 8, 3:44-50.

Tymoczko, T. ed.
    1985 *New Directions in the Philosophy of Mathematics.* Basel: Birkhauser.

Wilder, R. L.
    1968 *Evolution of Mathematical Concepts.* New York: John Wiley.

PART IV: MATHEMATICS, SOCIETY, AND SOCIAL CHANGE

11

Mathematics and Social Change

The message of this chapter is—in brief—the following: Human society today faces specific tasks which can be coped with only by additional *collective self-reflection*. Mathematics can be useful in fostering the process of self-reflection and hence in contributing to a beneficial development of humanity, if—and only if?—it undergoes some changes according to certain new orientations.

Essentially I am an optimist. I believe that in any case humanity will develop beneficially. By beneficially I mean that there will be less harm done to the individual, more working together rather than against each other, and more self-knowledge, although reverses and wrong directions cannot be avoided. In order to reduce such set backs and, above all, to make transitions in a way that people can accommodate to, we can do something. Further, I think that there is a danger that in the course of taking essential steps in human development, valuable goods are declared unimportant and therefore they are forgotten. Sometimes what they can contribute is not seen, or at least we do not see how important they can be for later tasks. I am especially interested in the mental goods of mathematics, the philosophy of which seems to contradict some observable and inevitable present-day tendencies in social life.

COMPLEXITY, SELF-ORGANIZATION, AND SELF-REGULATION

Before speaking about the potentials of mathematics, I want to indicate how I view the situation of human society today. I am not a social scientist, and therefore my diagnosis will be superficial, and moreover it must be based on the views of experts. The American social researcher John Naisbitt (1984) has pointed out ten so-called Megatrends, which he assumes will play an important role in the future of at least the Western hemisphere. (His method was simply to analyze more than two million articles in U.S. local papers.) Naisbitt discovered, among others, the following trends.

- a trend toward recognizing the necessity for strategic thinking in terms of big systems (temporal, spatial, etc.)—considering ecological, economic, technological and social networks;
- a trend toward decentralization—from hierarchy to social networks;
- a trend from either/or thinking toward a diversity of roles.

In a certain sense these trends seem to be a response to a phenomenon that is also frequently discussed today: *complexity*. Recently I had the opportunity to participate in a conference on "complexity and the human environment" (organized by the Honda Foundation in July 1987 in Vienna). Many contributions to that conference dealt with complexity in technological, economic, and ecological areas, and proposals were made for overcoming the problems: development of appropriate technologies, more (international) cooperation, generation of public consciousness about these problems, creation of a new ethic. In my view the contribution of the German sociologist Niklas Luhmann was of special importance. Through his deliberations, which were in a certain sense epistemological, he substantially enlarged the field under consideration. He said: Complexity is not only a property of some observed facts—for example, nature, technology or the economy—but also a *property of the observation*, meaning of the relationship between the "facts" and the observer. In order to understand the results of observations, the *observers must be observed* too. To the observation of "facts" by "society" the self-observation of the society itself must be added.

A first result of this self-observation and one possible explanation of complexity is the fact that society today is not a uniform, homogeneous block. Luhmann views it as a framework of relatively *autonomous social systems*. Examples of more or less autonomous social systems are the economy, scientific communities, and public administration. The notion of a social system seems to be of high importance in sociology today and for dealing with social development. In order to study social systems some borrowing of ideas from biology and cybernetics, developed there for explaining the nature and functioning of living organisms, has proved useful. The principle concepts are *'self-organization'* and *'self-reference,'* derived primarily from the biologists Umberto Maturana and Francisco Varela. In their view it is characteristic of living systems (in the biological sense) that they are able to reproduce their own organization (they call them *"autopoietic systems"*). Varela (1986) asks for the development of a systems theory that is more appropriate for living systems than classical systems theory, which follows the paradigm of dead machines. The main difference is that dead machine-systems can be described by their input/output relation, whereas for living systems their inner organization, their inner coherence, and their intrinsic structure are relevant. In living systems the interior system of rules—Varela calls it "operational closure"—leads to so-called eigen-behaviors, which are developed without influence from the environment. The interactions with the environment do not take the form of inputs into the organism (for example, information), which are processed, leading to outputs (observable behavior). Rather, self-organized systems select from the environment according to their interior organization, they interpret their environment, they give some things meaning and others not. Varela writes:

> for an autonomous machine characterized by its closure and its eigen-behavior, what happens is that these eigen-behaviors will specify out of the noise what of that noise is of relevance. So, what you have is a *laying down of a world*, a laying down of a relevant Umwelt. A world becomes specified or endowed with meaning; out of eigen-behaviors, there arises the possibility of generating sense. So what we are talking about here is the contrast between an instructive Turing automaton and an autonomous machine capable of creating (or generating) sense (Varela, 1986:119).

By the way, Varela arrives at a theory of evolution different from that which is based on the notion of optimal adaptation to the environment.

What is the relevance of these ideas from biology for social systems? It could be that the identity of even such systems has something to do with self-reference, operational closure, and the development of eigen-behaviors. This has always been assumed for most highly organized living beings, for humans. But Varela's deliberations rest on empirical findings concerning cells, nervous systems, or immune systems. If even these systems are able to create, through interactions of their parts, such a thing as collective spirit, shouldn't social systems be capable of doing this also? There are indications that this is the case, and this view has been adopted by some sociologists—as already mentioned—as well as by some management scientists. Without going into more detail now—I refer, for example, to Luhmann (1986) and Ulrich and Probst (1984)—I propose for further deliberation the assumption that the concept of a social system with the attributes self-organization and self-reference has relevance for human development.

In summing up our considerations so far, I would like to give three reasons why I think that self-reflection of social systems as well as systems of such systems is of special importance in the situation of society today:

1.   to obtain insight into the physical, biological, human "environment"; for example, in order to understand scientific propositions, one has to observe the scientists;
2.   to understand the identity and functioning of social systems, their organization and eigen-behaviors, to obtain knowledge of their hidden aims;
3.   to arrange, create, change such systems more consciously, and let them die if we wish.

To avoid misunderstandings: The self-reflection I mean is not an individualistic but a collective one, with mutual and self-observation about which it is necessary to communicate. Mutual observation is needed because of the blind spot that belongs to every system. The cybernetic scientist Heinz von Foerster speaks of a "second-order systematics."

INSTRUMENTALISM AND REFLECTION

Now the question arises: Can mathematics contribute to that process of collective self-reflection? At first glance there are opposed tendencies in mathematics. The most important such tendency is that mathematics is primarily viewed as an *instrument*, a tool, with which humans—as knowing or acting beings—confront themselves with a matter (for example, nature). With the help of this instrument they establish a *distance* between themselves and the matter, they express their differentness. As regards knowledge of nature, people have put mathematics between themselves and nature—though they belong to nature too. This kind of process counteracts self-reflection (cf. Fischer, 1993): see figure 6.2, p. 119.

Instrumentalism is the dominant paradigm of modern science, though by refutation of divine revelation and tradition, and by declaring human judgement the only source of knowing, the necessity of self-reflection is precisely rooted in modern science (see Hejl, 1984:61). It could be that humanity has not yet coped with the transition to modern science emotionally and therefore suppresses self-reflection.

Additional support for the view that mathematics is falsely oriented with respect to self-reflection of social systems arises from Varela's confrontation of machine-like and autonomous systems. As an instrument for the comprehension of input/output relations, as an aid in causing effects on systems from outside, mathematics has so far dealt more with machine-like systems. It is not by accident that mathematical systems theory is sometimes equated with "control theory."

But the *statements about the nature of mathematics* I made a moment ago *are themselves reflections*, not only reflections about a certain subject matter but also self-reflections about ourselves, about our relationship with mathematics and those matters to be handled through mathematics. More generally, each instrument that we use can be taken as an incentive, a starting point for self-reflection, for it mirrors properties, interests, and aims of people—individually and collectively. The more powerful an instrument is, the more potential for this mirroring it offers.

A second opportunity for using mathematics for social self-reflection comes from the fact that this reflection means a process of communication between people and groups of people. *This communication can be supported by mathematics.* In this case, of course, the instrumental character of mathematics is relevant, but with the aim of self-reflection, which influences the kind of tools as well as the handling of them.

In the following I want to propose *two orientations for mathematics* (as regards teaching, research, application), which correspond to the two above-mentioned potentials, and to describe these orientations in detail. They have been included within mathematics all along, but they have been concealed by crude instrumentalism. I call these orientations:

1. the analysis of basic assumptions of mathematical concepts and theories—mathematics as a mirror of humanity;

2.   problem description instead of problem solving—mathematics
as a means of presentation and communication.

ANALYSIS OF BASIC ASSUMPTIONS

Let me start with an example. The following formula is well
known:

$$K_t = K_0 \cdot q^t$$

It gives the amount of a capital K_t with compound interest after t
years, starting with K_0. Not so well known is the fact that this formula
is the solution of a system of functional equations. Taking K_t as a
function of K_0 and t, which means

$$K_t = f(K_0, t),$$

this system consists of the following equations (see Eichhorn 1978:10):

$$f(K_o + K_o^1, t) = f(K,t) + f(K_o^1, t)$$
$$f(K_o, t + t^1) = f(f(K_o,t),t^1).$$

Both equations can be interpreted as properties of *invariance with respect to splitting*. The first equation corresponds to a splitting of capital, the second to a splitting of time. More exactly, the first equation
says, if you split the original capital into two parts, say K_o and K_o^1 and
give them to—perhaps different—banks, you get the same result as if
you had invested the total amount at once. The second equation says,
if you invest a capital first for the time period t, then take the result
and invest it immediately for the time period t^1, you get the same as if
you had invested the original capital for the total period $t + t^1$. In a
certain sense these are the basic assumptions of compound interest
calculation. We know that they are not precisely fulfilled in reality.
You get a higher rate of interest for a larger capital as well as for an
investment over a longer time period.

An interpretation of this example in a social context could be the
following: The main purpose of large amounts of capital for long time
periods in our economy is *to control collaboration* in order to reach
more productivity than the sum of the productions of individuals
would yield. The classical formula for compound interest neglects this
purpose. It abstracts on the one hand from the effort necessary to

bring people—or capital that is the same in the economy—together and on the other hand from the fact that if you succeed in bringing people together you can get more than the sum of what can be achieved individually.

It is well known that mathematics itself is very much engaged in reflection on its concepts, theories, and procedures—in the above example, this is illustrated by proving the theorem that the formula for compound interest is the solution of a certain system of functional equations. One can say: The theories of applied and even of pure mathematics are the results of such reflections. Areas of study include examining the assumptions under which procedures can be applied, how the results can be interpreted, and so on (think, for example, of mathematical statistics). But there are *differences with respect to the relevance for social self-reflection.*

Take the concept of 'number.' One can view theories of the natural or of the real numbers as reflections about the concept of number. The focus is the logical dependence of statements, consistency, relative existence, or proof of nontrivial theorems (for example, concerning the distribution of primes). Another perspective on a reflection on the concept of number, which I view as more fundamental with respect to social self-reflection, is given by the *theory of measurement.* This theory is concerned with the question, under which assumptions is it meaningful to associate aspects of "reality" with numbers, and how can this be interpreted? A basic assumption of each measuring procedure is the construction of a mapping

$$f: A \to R$$

from a set A of "real" things into the real numbers, whereby certain relations in A are mapped onto "sensible" relations in the image set, for example, onto the natural ordering or onto addition. A simple theorem of measurement theory about the possibility of such a construction is the following:

THEOREM. *Let A be an at most countable set and S a binary relation in A. Then there exists a mapping $f:A \to R$ satisfying*

$$\forall a,b \ (\dashv A:aSb \leftrightarrow f(a) < f(b)$$

if and only if S is a strict weak order. (See Roberts, 1979: 101, 102, 109) (A strict weak order is an asymmetric relation the negation of which is transitive.)

According to the degree of uniqueness of the "measurement mapping" f different so called scale-types are defined, such as ratio scales, ordinal scales, and so forth. The theory yields the conclusion that certain operations with quantities are meaningful only under certain assumptions. To this belongs the well-known fact that the computation of the arithmetic mean is not meaningful for ordinal scales.

Though measurement theory remains within mathematical modes of inference and of representation, it also opens up fundamental perspectives. It studies the principles of what we are doing when measuring, but furthermore it poses questions about extensions and alternatives (for example, about nonnumerical or multidimensional measurement, cf. Roberts, 1979). Here I want to make a very concrete didactic suggestion: measurement theory should belong to the basic education of all mathematicians. Thereby the dominant tendency of theoretical refinement and of the search for interior dependencies with respect to one or more concepts would be supplemented by reflections about their relationship with things outside the world of mathematics, on a level that is more fundamental than treating only some isolated examples of model building.

Another example of a fertile field for reflection is *mathematical economics*. I have been occupied by this subject matter in some detail elsewhere (cf. Fischer 1987, 1988, 1991) and want to make only some general sugestions here. One of the basic assumptions of the so-called general equilibrium theory is the division of the economic world into different units. The relationships among these units are defined by the flow of commodities and of money and by information about prices. Additional interactions are neglected. Further it is assumed that each unit behaves "rationally," in the sense of maximizing a certain real function ("profit" or a given "utility function").

It has been argued that this theory is insufficient because it does not match reality. Kornai (1975), for instance, states that there is much more information exchanged between economic units besides price information. Going beyond this criticism I find it even dangerous because theories of social reality are not only descriptive but also prescriptive. That means, if such a theory is accepted people try to shape reality according to the theory. The theory yields an image of the economy as a *large machine*, with no collective will and freedom—but only will and freedom for the parts, namely the various units; they can, for example, choose their orders of preferences obeying some general consistency restrictions. Will and Freedom are divided; but so is rationality. The "economic man" who tries to maximize his profit or his utility function regardless of what others do is the central character. This concept of a 'divided, individualized rationality' inhibits coping with problems for which collective efforts are necessary, which are of another kind than the sum of the optimization efforts of all units. The classical game-theoretic example for such a situation is the "prisoner's dilemma" (Sen, 1982:63). Moreover, through a theory of the above kind a social system can be kept on the machine level, describable by input/output relations, and can be prevented from developing toward more consciousness and more freedom of action.

In a still more fundamental way the historian of economy Thomas Kuczynski (1987) sees limitations in a mathematization of economic matters. The fact that in set theory it is not permitted to define single elements with respect to the whole set—this leads to well-known antinomies—is, in his opinion, in contradiction to the comprehension of the individual as an "ensemble of social relations" (Fischer, 1988:20-28).

The final example for the orientation "mathematics as a mirror of humanity" I mention here is the most fundamental: considerations about *connections between the organization of knowledge and the organization of society.* The philosopher and social scientist Gerhard Schwarz (1985) interpreted the four axioms of *classical logic,* namely, identity, the exclusion of contradictions, the exclusion of the middle, and the principle of sufficient reason, as axioms of *hierarchy* as a mode of social organization. I have extended this interpretation by establishing the "equation" (Fischer, 1987):

logic : hierarchy = mathematics : bureaucracy.

The analogy is based upon various common aspects of mathematics and bureaucracy, such as autonomous rule system, materialization of the rules, and procedure orientation. To avoid misunderstandings, I should mention that I do not share the commonly held negative opinion about bureaucracy. For many matters I find it an efficient way of organizing social life; in any case, it was an historically necessary step in developing modes of organization. Studying and creating bureaucracies can also be very fascinating—as fascinating as studying and creating mathematics.

One can write the above equation also in the following way:

logic : mathematics = hierarchy : bureaucracy.

Now this means that logic plays a similar role for mathematics as does hierarchy for bureaucracy; both are in some sense the "organizational skeletons."

I recently read about a research result concerning the *problem-solving capacity of small groups* that indicates a connection between the kind of information processed and the organization of a group. If you have noiseless communication about clear matters, then hierarchically organized groups get better results (in coordinating information). But if "noise" is introduced by giving contents that cannot be spoken about so clearly, then "heterarchical" groups are better. They are able to develop a language, whereas hierarchical groups tend to fall apart (Foerster, 1984). Does this tell us anything about mathematics?

PROBLEM DESCRIPTION

Through the following example, taken from a lecture by Fred Roberts (1986), I shall demonstrate the difference between conventional mathematical problem-solving and what I propose. Roberts posed the following problem:

> We are given a rectangular street system of a city, each street being two-way. Because of heavy traffic there is frequent congestion and considerable air pollution. Therefore the government decides to make all streets one-way. How shall it be done?

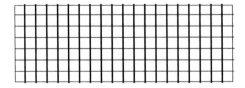

One postulate is of course the possibility of reaching every place from every other place. Mathematically formulated, the arising *digraph* must be strongly connected. But what other postulates are to be posed? What criteria exist for the efficiency of a solution? Roberts made the following proposals:

> For two vertices u,v of the street-system graph (that is, crossings) let $d(u,v)$ and $\bar{d}(u,v)$ be the lengths of the shortest connection in the old (undirected) and the new (directed) graph respectively. Consider the following quantities (n = number of vertices):

$$max_{v,v}\, \bar{d}\,(v,v) \tag{1}$$

$$\frac{1}{n^2-n}\ \Sigma_{v\neq v}\ \bar{d}\,(v,v) \tag{2}$$

$$max_{v,v}\,[\bar{d}\,(v,v)-d(v,v)] \tag{3}$$

$$\frac{1}{n^2-n}\ \Sigma_{v\neq v}\ [\bar{d}\,(v,v)-d(v,v)] \tag{4}$$

$$\frac{1}{n}\ \Sigma_v\ max_v\, \bar{d}\,(v,v) \tag{5}$$

$$^1\ \Sigma_v\ [max_v\, \bar{d}\,(v,v)\ ^-max_v\, \bar{d}\,(v,v)] \tag{6}$$

With respect to each of these quantities one can argue: The smaller this quantity the better the solution. An optimal solution is obtained if this quantity is minimal.

Roberts offered solutions for parts of the problem, especially for two of the above criteria for optimality, obtained by sophisticated combinatorial reasoning. This means he presented certain patterns for one-way street systems.

Now I pose the questions, *What does such a solution mean for practical purposes?* Not too much, I think. First, it is unlikely that all streets would be two-way and would have to be made one-way at once. And certainly also not all streets would be equally important. There are districts where people live and others where they work, the density of population need not be homogeneous, and so forth. This means that none of the given *"global" criteria* for optimality seems to be totally appropriate. Moreover, other global criteria could hardly do the job. Further, it is possible that the people of the city simply *want*, without clear reasons, a certain street to remain two-way. Finally, one has to take into consideration the public action of citizens if a "solution" were to be implemented in reality. Roberts made similar comments in his lecture.

Does this mean that mathematics is completely useless for this kind of situation? Not at all! I think that mathematical concepts and modes of representation are in this case appropriate for a *description of the problem.* Undirected and directed graphs offer a visualization of the situation and can be used for experiments (for example, simulations). The formulation of the quantities (1), (2), and (3) indicates different points of view on improving decisions. Further concepts can be developed, for example, considering the status of streets as main or side streets according to the present traffic, the system of public transport, the division into districts, and so on. For this purpose, concepts of graph theory can be used—for example, that of 'flow' in a weighted graph—or new ones can be invented. Of course the complexity of the problem would thereby increase, and a "mathematical" algorithmic solution would drift out of reach. (Even for criterion [1] it is a *n-p-*hard problem.) But it would *not* be the task of this process to find a final solution. Rather it would be to give a good description of the problem in order to enable the people of the city and their authorities to discuss the situation and make sensible decisions.

What about the connection with self-reflection? Clearly it is neither so fundamental as the relationship between mathematics and hierarchy, nor so fundamental as the deliberations about basic assumptions of the formula for compound interest. But it is the reflection on the needs and wishes of people in a certain social system, with respect to a given problem situation. Of course reflections of this kind are also included in conventional mathematical modellings. One wants to experience the interests and desires of the people concerned, in order to improve and refine the model. But the goal is always to obtain a (the) final solution. Therefore, those who construct the model are emotionally disturbed by the appearance of new aspects or contradictions. They want to be told once and for all what people want so they can concentrate on solving the problem. In contrast to this traditional orientation, the mathematician oriented according to my suggestion undertands him or herself as an *explainer of the problem* who helps people to articulate their imaginations, who points to alternatives—sometimes even as somebody who slows down the process of solving the problem. Whereas in the traditional way of proceeding the mathematician tries to reduce complexity and to exclude alternatives, in this he/she acts according to the "ethical imperative" formulated by Heinz von Foerster: "Act always so as to increase the number of choices." (1984:3). From various areas of application of mathematics voices can be perceived arguing in a similar direction as I am doing now—though mostly not as radically. In the following I want to briefly enter into three of those areas.

The first area is *applied systems analysis.* In the neighborhood of Vienna there is an "International Institute for Applied Systems Analysis," a joint institution of Western and Eastern industrial nations. In one of the projects at the Institute, computer simulation models are constructed for large ecological, economic, and other systems. The director of the institute, Thomas H. Lee, at the conference on complexities and the human environment, which I have already mentioned, spoke about a fundamental change in the modelling of such systems. Instead of the use of classical mathematical methods, which provide solutions, in recent years so-called interactive decision-support systems have been developed. The task of these systems is not to relieve people of decision making but to support them in the process of reaching their decisions. This orientation has been expressed (even more articulately) by the well-known systems analyst Dennis Meadows in a lecture in Vienna: The task of so-called strategic models is not to predict the future but rather to accomplish the following functions (Meadows, 1986:83).

- research
- communication
- legitimation
- education.

Only the first of these four functions is primarily related to the "matter": research means exploration of the problem situation. The other three tasks have explicitly to do with communication and the construction of social relations ("legitimation" means to validate the status of the model constructor among socially relevant persons with respect to the problem).

A second area where the tendency towards mathematics for problem description can be seen is *mathematical economics*. One of the important problems in economics is to judge economic inequality (in a nation, for example). Several measures for inequality have been developed by economists. I shall present two of them for illustrative purposes. For the first, let $y_1, y_2, \ldots, y_n$ be the incomes of the persons in a community and μ the arithmetic mean of these incomes. Then

$$C = \frac{\sqrt{\frac{1}{n} \sum_{i=1}^{n} (y_i - \mu)^2}}{u}$$

is called the "coefficient of variation" of this income distribution (Sen, 1973:27). For the second measure we need a function that associates to each income the utility to be drawn out of this income. Such a function U ("utility function") is generally assumed to be increasing and concave, such that with increasing income less and less utility can be drawn from one unit of money. Now the following measure of inequality can be established.

$$D = 1 - \sum_{i=1}^{n} \frac{U(y_i)}{n \cdot U(\mu)}$$

This is called "Dalton's measure" (cf. Sen, 1973:37). If U is *strongly* concave, then $D = 0$ if and only if all y_1 are equal. For C, the corresponding property is clear immediately. For both measures it is feasible: the larger the inequality, the larger the corresponding measure. But the measures are different. Ordering nations according to their "amount" of inequality of income distribution can yield different orders of succession for different measures. Furthermore, the measures have different properties that are considered as more or less desirable. For example, it is interesting how a measure changes when an income transfer is made from the rich to the poor in a range of middle or of high income. (Of course in any case the measure should become smaller.) Or, should an inequality measure take into consideration the absolute income level; should it remain invariant when all incomes are changed proportionally or additively? A basic difference is seen between "descriptive" measures such as the coefficient of variation, and so-called normative measures, such as Dalton's, where a certain "norm" is introduced by the utility function. The economist Amartya Sen pleads for a simultaneous use of the different measures and argues that the individual measures should not be interpreted too strongly. He writes:

> I have tried to argue in favour of weakening the inequality measures in more than one sense. First of all, the mixture of partly descriptive and partly normative considerations weakens the purity of an inequality index. A purely descriptive measure lacks motivation, while a purely normative measure seems to miss important features of the concept of inequality . . . Second, even as normative indicators the inequality measures are best viewed as "non-compulsive" judgments recommending something but not with absolutely compelling force. This has implications in terms of the treatment of inequality rankings as prima facie arguments and permitting situation-specific considerations to be brought into the evaluation if such supplementation is needed (Sen, 1973:75).

Obviously this means dealing with mathematics in a way that corresponds to the idea of presentation, communication, and description of problems in the sense of collective self-reflection.

As a third area where the described tendency can be realized—a tendency which of course has always existed but which in my view is now coming more explicitly into the minds of researchers—I take the *applications of mathematics in psychology and sociology.* I shall only quote the psychologist Lee Cronbach, who is to be credited with the introduction of mathematical methods in psychology and who has helped stimulate a boom of research results according to the pattern of "aptitude-treatment interactions" in the U.S., for which mathematical statistics is an inevitable tool (for more details see Fischer, 1987). He is today professor emeritus at Stanford University, and for more than ten years he has written articles critical of the direction of the development of the social sciences. Though the following is not directly concerned with the use of mathematical methods, I think that conclusions can be drawn for mathematics:

> Concepts contribute to pluralistic decision making by helping participants examine their situations and values . . . The social scientist helps not by playing expert but by playing educator, eternally pressing the question, "Have you taken X and Z into account?" Social science is cumulative not in possessing ever-more-refined answers about fixed questions but in possessing an ever-richer repertoire of questions. The educative influence of a piece of research may extend far into the future. Concepts have enduring value, and so does a sense of what-connects-to-what (Cronbach, 1982:72-73).

DISCIPLINING AND OBJECTIVIZATION

The two orientations "mathematics as a mirror of humanity" and "mathematics as a means of communication" are concerned with analysis and synthesis of social reality. The first orientation gives more emphasis to analysis. By reflecting about given mathematics one tries to *analyze social relations* that are mirrored in this mathematics. The second orientation is more directed towards synthesis. Through communication, *social relations are constructed*, social systems with their identity, operational closure, and eigen-behavior are established. Social systems have always been constructed with the help of mathematics. Pedagogues have called the effect of this process on the individual "disciplining" (*Disziplinierung* in German, meaning "shaping people like soldiers") and have criticized it. I think that disciplining is inevitable when constructing social systems; I think that even more

212 *Mathematics, Society, and Social Change*

disciplining by mathematics will be needed in the future, but that this should be done more consciously and that the means should be more appropriate to the aim of constructing systems, the elements of which are humans.

The function of mathematics in the process of collective self-reflection is objectivization in the sense of reification. Social and other relations are made objects, and I mean this also literally in the sense of materialization. Mathematics offers visual pictures, which usually have to be interpreted in a symbolic way; that means that to understand them one has to know certain social conventions. Pictures of this kind are indispensible for communication if it is to be expanded beyond the scope of small groups. It is not the case that we should use mathematics for communicative self-reflection simply because it is here—though this would be a sufficient reason—but without mathematics or, better, without a world of symbols reflecting certain relations, we would be in trouble. The French prehistorian and anthropologist André Leroi-Gourhan even writes, A "society with a declining capacity for creating symbols would lose its capacity for action" (1980:267, my translation; see also Fischer, 1984). In this spirit, and considering the potentials of computers, I believe that there will be a shift of importance in mathematics from handling given symbol systems toward the *creation of new symbols*, in connection with studying the suitability and social relevance of the symbols. Maybe this is simultaneously a step towards a "demystification" of the symbols. There is always the danger that a necessary objectivization of the abstract leads by mystification to a *Verabsolutierung* (making absolute) of present states. By the way, a very powerful, and in a wide sense mathematical, objectivization of abstract issues is *money*—where we face the cited danger every day.

It would be interesting to analyze the *relationship between disciplining and objectivization* through mathematics from a sociophilosophical point of view. Objectivization of social relations on the one hand means to establish a distance between ourselves and exactly those relations; thereby it is a means to get free of them. We become able to handle, to manipulate these relations—all this can increase freedom of the individual and of society. On the other hand, objectivization gives the social relations an absolute, maybe even eternal character—thus disciplining gets more emphasis, it becomes more severe, rigid, and oppressive. The construction of social systems with mathematics is pursued in such a way that there is only one possibility: nobody can escape. The "insight into the necessity" or the lack of it can make things repressive for the individual.

I think the way to keep the advantages and to avoid the disadvantages of objectivization is to *make the process of objectivization*, the construction of symbols, *more explicit* and to make it available to people. Thus the objects would lose their absolute, eternal character; it would be seen that there are alternatives. Simultaneously the *construction of social systems and the inevitable disciplining would become more playful*, as is the case for the construction of mathematical systems according to a formalist philosophy of mathematics. I think that one of the problems of social change is that we are too serious about social matters. This seriousness does not allow the flexibility we would need in order to cope with the problems of today. Mathematics is also guilty of this seriousness. Mathematics would also offer play and freedom. To state it as a prophecy: Socially we have not yet attained that degree of freedom and flexibility which we have reached in pure mathematics, but as an optimist, I say, We are on the way.

MANAGEMENT—HOW REFLECTION BECOMES AN INSTRUMENT

My espousal of more reflection looks easy, if done from the position of a philosopher. What happens, however, if one stands "in reality," if one is to bring about something, if one is responsible for a social system? Usually the task of inducing social change is assigned to *management*. Since World War II mathematization has taken place in this field under the title "operations research." In recent years a direction of discussion has gained importance among theorists of management that on the one hand points to the limitations of mathematical methods and that on the other shares a view of social systems I described earlier. The two Swiss scientists in the field of business administration, Fredmund Malik and Gilbert Probst (1984), describe essential features of this view under the title "Evolutionary Management." I want to deal briefly with three characteristics of this kind of management and with their relationship to mathematics.

The first characteristic is the comprehension of even *artificial, human-constructed organizations*, such as firms, as *self-organized, self-developing social systems*. That means, for example, that a firm is not determined only by the purposes that have been formulated officially or by the structures that have been created consciously. Rather it is a social body that has its own identity, developed by more or less conscious interactions. This social body partially further functions according to hidden rules; it is therefore, in a certain sense, incalculable.

Human actions, but not always human intentions, are causes for developments. Without conscious will, rules can arise depending on which matters are processed. As a consequence we get rational planning and acting that is limited narrowly; in particular, conventional mathematical methods oriented towards "solutions" cannot grasp essential areas of management.

The second point is the consistent view of *the manager as being part of the system*. The manager must be part of the system if he or she wants to bring anything about. In particular, managers must enter into relations which go beyond the gathering of objective information. If they do, they may be in a position to influence processes—as catalysts, so to speak. Hubert Dreyfus and Stuart Dreyfus, computer critics, have devoted one chapter to "Managerial Art and Management Science" in their book *Mind Over Machine: The Power of Human Intuition and Expertise in the Era of the Computer* (1986). There they criticize, on the basis of some examples, the use of mathematical methods in management. The tenor of their criticism is that mathematical methods take account of only context-free, dissected information. Using these methods, the manager is able to reach only a very low level on the way to becoming an expert, namely that of "advanced beginner" or of "competence." More advanced stages like "proficiency" and "expertise" are attainable only by an intuitive connection with the respective social system. The method of dissociated information processing corresponds, in their opinion, to a style of career with frequent change of jobs ("job-hopping"), which was, and may still be, very common in the U.S. They think that the success of Japanese companies has to do with the tight connection between the managers and the social systems in which they work. About the mathematical methods they write:

> In general, all formal models of decision-making ask the expert questions which place him in a detached objective position and so fail to tap his intuitive expertise. They suffer, just as does conventional AI and expert systems engineering, from the impossibility of replacing involved knowing how with detached knowing that. The same problem reappears in attempts to automate factories and offices. (Dreyfus and Dreyfus, 1986:184)

That does not mean that the authors would reject mathematical methods totally. For example, they suppose them to be useful as "decision-support systems" in the sense of "what if" models. Moreover they mention recently developed computer aids for supporting communication and even conversation among managers because "decision-making is only one part of a manager's activity." They write

> For example, "coordinators" is the name given to a new family of microcomputer tools developed by Fernando Flores of Action Technologies, a San Francisco company, Terry Winograd of Stanford University, and others. Their idea is that managerial action is necessarily social and involves people interacting verbally. Coordinators are designed to facilitate actions of people working with each other. Using a coordinator the manager conducts his business by inquiring, instructing, ordering, questioning, requesting, proposing, inviting, promising, and reporting. As he does, the coordinator automatically sends his message to others and elicits communication from others. Of course, the computer has no understanding of the messages it is sending. Still, by keeping track of them the tool facilitates the ongoing conversations that are essential to management (Dreyfus and Dreyfus, 1986:192).

The authors distinguish these tools from the so-called management-information systems which often yield only a clumsy abundance of data.

The third characteristic of the described new direction in management consists in the view that an experienced manager makes decisions not on the basis of conscious deliberations about the future but on the basis of a *sound knowledge of the present*. It is not a sophisticated prediction model that is decisive, but an intuitive comprehension of "what is" (cf. Dreyfus and Dreyfus, 1986). How then can the manager bring something about? By confronting a social system with its present reality, is an answer that is given by Stafford Beer (1986), a management consultant and cybernetician. For him this is an element of the Asiatic way of thinking and art of managing. In the Western hemisphere one tries to define goals and then to get an organization there, an effort that often fails. The other possibility is to confront a system with its structures, needs, and wishes, to let it define its goals and then to use its inner dynamics to let it reach them—with some help and guidance. Self-reflections thus become an instrument—and indeed an absolutely efficient one, if one believes the reports of various management consultants.

A final remark: a fundamental problem of social change today is—and possibly has always been—the handling of the conflicting tendencies *interweaving* and *autonomy*. On the one hand the task is to establish connections and become aware of connections; on the other hand a desire and need for independence also exists. Mathematics is able to contribute in various ways. Its contribution in becoming aware of connections is well known. That mathematics itself can be an expression of social relations and structures and that it can create them, is less obvious. But precisely by reflecting these facts, through its function as a mirror of social affairs, it can contribute to autonomy. Simultaneously mathematics is able, through a better understanding of its role as a means of communication, to create new opportunities for connections between people and social systems. Thus it can again contribute to interweaving, but on a new level of quality. We should not permit mathematics to be totally occupied by its traditional tasks.

NOTE

Extended version of a lecture presented at the 3rd International Conference on the Teaching of Mathematical Modelling and Applications (ICTMA) in September 1987 in Kassel, FRG. I am indebted to W. Nemser for going over a preliminary draft eliminating language mistakes and improving style.

REFERENCES

Bammé, A.
  1986   "Wenn aus Chaos Ordnung wird: Die Herausforderung der Sozial-
         wissenschaften durch Naturwissenschaftler." University of Klagen-
         furt. Manuscript.

Beer, S.
  1986   "Recursions of Power." Pp. 3-17 in Trappl (1986).

Cronbach, L. J.
  1982   "Prudent Aspirations for Social Inquiry." In W. H. Kruskal, (ed.), *The
         Social Sciences: Their Nature and Uses.* Chicago: University of Chicago
         Press.

Dorfler, W., R. Fischer, W. Peschek, eds.
  1987   *Wirtschaftsmathematik in Beruf und Ausbildung.* Vortage beim 5. Karnt-
         ner Symposium fur Didaktik der Mathematik. Vienna: Hölder-Pich-
         ler-Tempsky; Stuttgart: B. G. Teubner.

Dreyfus, H. L., and S. E. Dreyfus
  1986   *Mind Over Machine: The Power of Human Intuition and Expertise in the
         Era of the Computer.* New York: Free Press.

Eichhorn, W.
1978 *Functional Equations in Economics.* Reading, Mass.: Addison-Wesley.

Fischer, R.
1984 "Offene Mathematik und Visualisierung." *Mathematica Didactica* 7:139-60.
1987 "Mathematik und gesellschaftlicher Wandel." Projektbericht im Auftrag des Bundesministeriums für Wissenschaft und Forschung. University of Klagenfurt.
1993 "Mathematics as a Means and as a System." Chap. 6 in this volume.

Fischer, R., G. Malle, with H. Burger
1985 *Mensch und Mathematik: Eine Einführung in didaktisches Denken und Handeln.* Mannheim: Bibliographisches Institut.

Foerster, H. von
1984 "Principles of Self-Organization—In a Socio-Managerial Context." Pp. 2-24 in Ulrich and Probst (1984).

Hejl, P. M.
1984 "Towards a Theory of Social Systems: Self-Organization and Self-Maintenance, Self-Reference and Syn-Reference." Pp. 60-78 in Ulrich and Probst (1984).

Kornai, J.
1975 *Anti Äquilibirium: Über die Theorien der Wirtschaftssysteme und die damit verbundenen Forschungsaufgaben.* Berlin, Heidelberg, New York: Springer.

Kuczynski, T.
1987 "Einige Überlegungen zur Entwicklung der Beziehungen zwischen Mathematik und Wirtschaft (unter besonderer Berücksichtigung der Entwicklung der Wirtschaftsmathematik)." In Dorfler, Fischer, and Peschek (1987).

Leroi-Gourhan, A.
1980 *Hand und Wort.* Frankfurt: Suhrkamp.

Luhmann, N.
1986 *Okologische Kommunikation.* Opladen: Westdeutscher Verlag.

Malik, F., and G. J. B. Probst
1984 "Evolutionary Management." Pp. 105-20 in Ulrich and Probst (1984).

Meadows, D. L.
1986 "Guidelines for Influencing Social Policy through Strategic Computer Simulation Models." Pp. 81-95 in Trappl (1986).

Naisbitt, J.
1984 *Megatrends: 10 Perspektiven, die unser Leben verandern werden.* Bayreuth: Hestia.

Narens, L., and D. R. Luce
   1986 "Measurement: The Theory of Numerical Assignments." *Psychological Bulletin* 99, 2:166-80.

Roberts, F. S.
   1979 *Measurement: Theory with Applications to Decisionmaking, Utility and the Social Sciences.* Reading, Mass. Addison-Wesley.
   1986 "Applications of Discrete Mathematics." Lectures at the Claremont Graduate School, California, in June.

Schwarz, G.
   1985 *Die "Heilige Ordnung der Männer": Patriarchalische Hierarchie und Gruppendynamik.* Opladen: Westdeutscher Verlag.

Sen, A.
   1973 *On Economic Inequality.* Oxford: Clarendon Press.
   1982 *Choice, Welfare and Measurement.* Cambridge, Mass.: MIT Press.

Trappl, R., ed.
   1986 *Power, Autonomy, Utopia: New Approaches towards Complex Systems.* New York and London: Plenum Press.

Ulrich, I., G. J. B. Probst, eds.
   1984 *Self-Organization and Management of Social Systems.* Berlin: Springer.

Varela, F.
   1986 "Steps to a Cybernetics of Autonomy." Pp. 117-22 in Trappl (1986).

12

The Social System of Mathematics and National Socialism: A Survey

Studies of the history of scientific disciplines in Nazi Germany have largely concentrated on the intrusion of Nazi ideology into scientific thinking (e.g., Beyerchen, 1977; Mehrtens and Richter, 1980). Few works have addressed the process of the integration of a science into the society (cf. the excellent book of Geuter, 1984). As a result, there has been no serious attempt to work out a comprehensive vision of the position of a science in German society and its development in the thirties and forties. Such a vision must include not only the core of scientific research and teaching, but also the position in the general system of education; the fields of application; the neighboring scientific and technical disciplines; the political relations of the discipline; and the situation of individual scientists and their relation to the disciplinary system. In brief, it has to comprehend the complete system of the discipline and all its essential relations to its environment. For careful, fact-oriented historiography this would be an almost insurmountable problem. But it is also a problem of theory. What is the "complete social system" of a science? How does it work? What are "essential" environment relations? The basic understanding of a "social system" used here follows from Luhmann (1984). It includes as essential an interest in the internal organization of the system, environment-system relations (where individuals are considered part of the environment), and the processes of identity formation and boundary maintenance.

THE SOCIAL SYSTEM OF MATHEMATICS

During the nineteenth century, mathematics in Germany became an institutionalized, fully professional scientific discipline with a well-established and central position in the systems of knowledge production and education. "Mathematics" means here the social system of the discipline with its central function of producing and disseminating novel knowledge of a specific character. As two of the most important characteristics of mathematical knowledge, one might point out its pervasive role as a linguistic and theoretical tool in the production of scientific knowledge and the fact that it has no well-defined object in the material world and thus no clear boundaries to its applications. Further, the extremely compelling character of mathematical argument must be noted. Like other knowledge-producing social systems, mathematics defines its identity through the specificities of its knowledge. Because of the constitutive character of identity and difference (from other systems) the system has developed a core in which "pure" mathematics is produced. "Purity" thus serves an extremely important function by producing and reproducing the identity of the system and important elements of the identity of individuals taking part in the system.

Like other systems, mathematics has to care for productivity, social legitimacy, and professional autonomy if it is to survive. These functions have to be fulfilled in order to reproduce the internal coherence of the system but even more so to ensure stable environmental relations. Stable relations with the environment are important to intellectual endeavors such as mathematics, since they do not produce the material resources necessary for reproduction, but must depend upon others for them. The basic thesis presented here is that a scientific discipline exchanges its knowledge products plus political loyalty in return for material resources plus social legitimacy. Obviously such a trade is the result of continual bargaining and is subject to historical change. The systematic analysis would have to consider among other points the position of the discipline in the larger system of the sciences, the power of experts in politics or the economy, and the role of instrumental rationality in modern societies.

To ensure productivity the system has to care for recruitment of personnel; it has to give room for some social and cognitive deviance in order to maintain creativity; and both internal and external communication have to be organized. Organized communication appears to be a most important characteristic of knowledge-producing social systems. Consequently the organization of journals, professional societies, and conferences play a central role. Furthermore a certain degree of social and cognitive coherence within the system is a precondition of sufficiently intense communication. The latter is produced by what has been called the "paradigm" or "disciplinary matrix" of the field (Kuhn, 1971). The system itself is highly oriented towards knowledge that reinforces its productivity. Similarly the exchange of rewards, or the ways to gain scientific "capital" (Bourdieu, 1975), is also related to notions of productivity.

Social legitimacy for mathematics is basically attained in two ways; through the utility of mathematics, and through its cultural value. Cultural value can range from metaphors like "honor of the human spirit" to "expression of the German character," or from "basic training of the mind" to "core of modern rationality." Since the utility of novelties in pure mathematics is usually not immediately visible (nor is its cultural value), mathematicians have frequently been defensive, complaining about the lack of understanding for their field. Social legitimacy has to be such that the basic and supporting institutions are preserved and possibly extended. Basic institutions are those where mathematical production takes place. In Germany this means almost exclusively universities and, since the end of the nineteenth century, institutes of technology. Supporting institutions would be places where mathematics is applied and taught. In the German system legitimacy of an academic subject that has no immediate field of application in society rests to a very large extent on the status of the subject in secondary education, since teacher education and scholarly production are combined in the universities. For this reason, school mathematics is a field of legitimation of extreme importance in Germany. There is the constant interest to maintain the identification of mathematics teachers with the academic subject. And in this field the cultural value of mathematics has been important, because in general the mathematics taught in schools appear to be of little use to the students. The utility of mathematics on the other hand is strongly expressed by its integration in the system of natural sciences and technology. Again a certain amount of identification with mathematics within other social systems has to be produced. The fact that mathematics has no defined field of application in the material world has the effect that strategies of legitimation are strongly mediated (for example, directed at other professions like teachers or natural scientists). Within this structure the role and development of "applied mathematics" as a distinct subfield is of high interest.

The professional autonomy of mathematics is ensured mainly by monopolizing competence for the specific type of knowledge production. In the discussion of the modernization of mathematics below, I shall try to show how the development of knowledge in its basic characteristic interacts with its social function in ensuring autonomy. A dense and far-reaching (international) communication network within the system serves the goal of autonomy by monopolizing competence and means of communication but also via its productive function. A social system that depends on international communication for its central functions (production) gains autonomy in the face of national powers. On the other hand material resources come from national powers. Thus a dynamic equilibrium between autonomy and integration in the national society is necessary. A similar equilibrium is needed between utility and autonomy for legitimacy, since the competence to decide what mathematics is taught and where it is applied does not rest in the hands of the mathematicians. In times when the cultural value of the field is in doubt, the move towards a utility argument is a delicate problem.

This rough sketch of some basics of the social system of mathematics must suffice as a basis for the following discussions. It should have indicated that the system has necessary inner tensions and is of a highly dynamic stability.

ORIGINS AND POWER STRUCTURES OF THE THIRD REICH

The following remarks are solely meant as a background to the specific theme of this paper. I can merely indicate a few basic problems that are still matters of intense discussion. But first of all two popular misconceptions have to be removed. Firstly, the ideology of national socialism cannot be reduced to "blood and soil." The ideology is a large collection of pieces, some of which were decisive for political action while others were the object of cynical manipulation. This collection was hardly coherent as a cognitive system. Instead the parts could be combined and recombined as necessary into what may be called "regional ideologies." The concept of a "German" mathematics or physics was such a regional ideology that related to some core elements of general ideology but in a specific way, adapted to the social and cognitive situation of the discipline. Nazi ideology also embraced a cult of technics and naturalistic scientistic modes of justification. Thus regional ideologies in technology or in chemistry could be constructed along the lines of power, leadership, German imperialism and superiority, but excluding or at least playing down anti-Semitism, the cult of the soil and of rural life, and romantic irrationalism.

The "reactionary modernism" in Germany, as part of Nazi ideology, has recently been discussed by Herf (1984).

Secondly the notion of a totalitarian, tightly structured and controlled state has to be revised. Power struggles were common on all levels below the führer, and these conflicts included interests from outside the party. Further, the system of Nazi society was borne mainly by four blocks of power: the party, state-bureaucracy, industry, and the army. These powers cooperated and competed. Thus the environment for a social system like mathematics was much more diversified and dynamic than appears on first sight. Adaptive moves could relate to various sources of power. The immediate link to the state lay in university administration, and here the bureaucracy of the ministry of education was involved in constant conflicts with party agencies. Systems like mathematics obviously preferred the bureaucratic rationality of ministerial administration. Similarly industry was a potential partner. The physicists' association, for example, elected an industrial physicist as president explicitly in order to use industrial power to counter party influence.

An important theory for the explanation of the rise of fascism and especially national socialism is to see it as a reaction to the problems of modernization. It is argued that rapid industrialization, urbanization, rationalization, and bureaucratization in Germany did not find equivalent changes in politics, culture, and mentality. National socialism arose as an antimodernist revolt mainly of those individuals who were in fear of, or actually subject to, social marginalization. In its effect, the argument goes, national socialism modernized German society by destroying traditional institutions. The whole argument is subject to varied discussions. But it is, as an explanation of the rise of German fascism, and of its large support, at least partially valid. It is confirmed by the history of the sciences, but especially in this case needs a careful consideration of national socialist modernism (cf. Herf, 1984) and the role of technocracy in the power structures of the Third Reich.

Finally the Nazi system was changing during its twelve-year reign. For the history of mathematics and the natural sciences the years 1936-37 mark an important change. With the four-year-plan and the intensification of war preparations several institutions strengthened the position of technocratic rationality to which social systems that produced technically applicable knowledge could relate. Another such change occurred in 1942-43, when the time of quick victories was over, and even party agencies that stubbornly had insisted on the value of a romanticist "German" science dropped this notion in favor of technological and scientific rationality.

MODERNISM, ANTIMODERNISM, AND NAZI IDEOLOGY IN MATHEMATICS

By 1930 it was quite clear what the term *modern* meant when applied to mathematics: the conceptual study of abstract mathematical concepts characterized by axioms valid for sets of otherwise undefined elements and presented by proceeding from the elementary concepts to the more complicated structures; a hierarchical system of mathematical truths rigorously proved, the language applied having hardly any other function than to label the objects and to ensure the validity of statements. "Modern" mathematics in this sense had no extramathematical meaning, did not indicate possible fields or objects of application, was devoid of hints to the historical or heuristical background of the theory, and at most was in a highly implicit manner structured along didactical guidelines.

The thesis presented here is that mathematical modernism emerged slowly through the nineteenth century, became fully visible around the turn of the century, and gained its position as a dominant and productive research program by the end of the 1920s. By then antimodernist positions resided without influence in marginalized fields and groups. Attacks at the social and cognitive dominance of modernism occurred with the help of powers outside the discipline, as in Ludwig Bieberbach's attempt to establish a "German" mathematics with the help of national socialism. For the analysis of the latter development, a thorough analysis of the genesis and the structure of modernism and traditionalism in mathematics would be necessary.

Sociologically it is crucial for such an analysis to understand the close interrelation of social and cognitive modernization of the discipline and the way this is embedded in the environment of mathematics. The key concepts are 'autonomy' (both cognitive and social) and 'universality.' Mathematics gained cognitive autonomy during the nineteenth century through deontologization, self-justification, and productive self-reference. Deontologization meant giving up the basic assumption that mathematics is concerned with real space and with magnitudes as applied in the real world. The legitimacy lost through giving up ontological ties to the real world was compensated through self-justification by method. Instead of the inquiry into the "nature of the object" it was the methodologically and logically controlled conceptualization of relations and operations which became the object and justification of mathematical knowledge. With this mathematics became, in its "pure" core of knowledge production, self-referential. Problems, objects, and solution criteria were produced and controlled by mathematical thinking.

All this became visible most clearly in the rise of set theory, which on the one hand made it possible to talk and argue about objects that have no other characteristic than being capable of relations and operations. On the other hand, set theory gave access to the formal handling of infinity. A complementary event was the rise of mathematical logic and the axiomatic method that allowed for a rigid control of the syntax of mathematical expression and argument.

All this adds up to cognitive autonomy insofar as no other approach or field of knowledge can intervene in the basics of mathematics unless it becomes itself part of mathematics. No specific field of objects or relations or operations is presupposed. Anything can become the object of mathematical thought. At the same time mathematics is rigid and compelling. Accept the methods and the axioms and there is no way to evade the consequences. Thus mathematics is cognitively universal both in its procedure and in its applicability.

As implied above, cognitive autonomy implies social autonomy to some extent. It means potentially complete monopoly of competence for a certain field of mental activities. Internally the rigid control of the syntax of mathematical expression and the tendency to push out various semantic dimensions from formalized mathematical discourse is oriented extremely toward the professional production of novel knowledge. The rigid discrimination of formal and informal discourse produces a highly esoteric core, inaccessible to outsiders. The specific structure of cognitive autonomy and universality is also productive in that it allows high interpenetrability of mathematical subfields, thus generating new problems and solutions. Given that mathematics is applicable and is used by others, the discipline is also socially universal. There are no intrinsic limitations to where mathematics may find its market. Any society, any social system, any other system of knowledge production can buy mathematical results and competence to use it as needed.

Obviously modern mathematics has to complement its core autonomy with structural measures to ensure that there is demand for mathematics, either as a tool to be applied or as a paradigm for the prevalent type of rationality. Consequently, the social system of mathematics needs internal task-oriented subdivisions and institutionalized ties to its environment. Around the turn of the century, when modern mathematics became prominent, such task-oriented subsystems came into appearance. One was the institutionally emerging specialty of applied mathematics with a specific identity; the other, the field of a pedagogy of teacher education in mathematics and the relation between school and university mathematics. Both developed partly from outside pressures from teachers, engineers, natural scientists, and industrialists. The central cognitive definition of mathematics by its pure, modern core was related to those task-oriented fields. Lectures on "elementary mathematics from higher standpoint" served, for example, to relate the pedagogical branch to the development of scientific mathematics, and to attempt to axiomatize mathematical theories for physics aimed at an immediate relation between this field of application and pure mathematics. Applied mathematics, institutionalized with a wide range of very different tasks, had problems achieving coherence because there was no central governing theory or methodology that would cover most of the fields, from numerical analysis and graphical statics to mechanics and mathematical statistics.

This description is obviously idealized. The history is much more complicated. The relations indicated between various parts and functions of the system should not be taken as indicating historical causation. It should, on this level, be understood as the self-organization of a social system interacting with its environment. Further, it must be stressed that mathematics was embedded in the larger system of the sciences that fulfilled important functions for all subfields, for example, the legitimation of scientific rationality.

Modernization in mathematics, as elsewhere, meant losses. The "Loss of Certainty" (Kline, 1980) occurred cognitively as well as socially, because autonomy and universality imply that values and standards of action are no longer incorporated but rather a matter of implicit negotiation. Similarly, the sharper demarcations between, for example, formal and informal mathematical discourse, pure and applied mathematics, mathematics and physics imply a loss of connectedness of the whole system and its environment. Finally the modern mathematics, especially in formal discourse, no longer carried any meaning other than that for mathematical research work, and even there the heuristic context was obscured. As could be expected modernization was criticized by traditionalists; the criticisms sharpened when the paradoxes of set theory emerged around 1900. The critique centered on the loss of meaning. It was said that formalization and logic in mathematics had gone too far. Parts of mathematics became senseless and were uncontrollable because they could not be constructed out of elementary thought-action (like counting). Such mathematics no longer had any meaning in the real world. Further, "intuition" (in various forms) was defended as a guideline, tool, and justification of mathematics. This type of attack was frequently combined with the argument that modern mathematics would hinder the educational and applied tasks, while the establishment of a separate applied mathematics would destroy the unity of the discipline.

Ludwig Bieberbach, the later champion of "German" mathematics, used all these arguments during the twenties. In a lecture in 1926, for example, he sharply attacked Hilbert, the dean of mathematical modernity, and depicted the modern theories as "skeletons in the sand of the desert of which nobody knows whence they come and what they have served for." Bieberbach was a mathematician of high reputation, well established and with a strong position in the German mathematical system. In his work, however, he considered himself as a "geometrical thinker," relying on intuition and mental imagery. With this emphasis he had a tendency to be pushed to the margin of the official and progressive mathematics. He was a pure mathematician with strong interests in application, fearing a new field of applied mathematics as a "wedge between mathematics and its users." To him, the danger of modernity centered on both the loss of meaning, by extreme formalization, and the loss of unity, through differentiation.

To these combined motives for antimodernism were added a series of conflicts, in which Bieberbach sided against the dominant position held by the mathematicians of the Göttingen Institute. Bieberbach sided with L. E. J. Brouwer's intuitionism and fought against David Hilbert's formalist program. This struggle had strong elements of competition for social power in the discipline. Hilbert's victory was obvious when he succeeded in eliminating Brouwer, and Bieberbach with him, from the editorial board of the influential journal *Mathematische Annalen*. Bieberbach was involved in a series of further small conflicts that placed him in clear opposition to the dominant (modernist) Göttingen school. Most important for the present analysis was the combined effort of Bieberbach and Brouwer to organize a German boycott of the International Mathematics Congress of 1928. Previous congresses after World War I had excluded the Germans, and now nationalists called for a counterboycott. These actions were strongly opposed by Hilbert, who eventually led a large German group that attended the Congress. The obvious problem of interpretation is to understand the connection between political nationalism and the specific philosophical position in mathematics. Was there a systematic reason for this relation? Was intuitionism even prefascist? I shall return to these questions after briefly sketching Bieberbach's Nazi activities.

To the surprise of his colleagues and students, Bieberbach, early in 1933, turned to the Nazis. Besides several Nazi-related activities at the university and the academy he made his reputation as *the* Nazi among mathematicians by his theories on the psychological (and thus racial) background of different mathematical styles, which he published in lectures and articles beginning in 1934. He was fairly careful not to express anti-Semitic or nationalist prejudices too bluntly, but gave an all too obvious basis to draw such conclusions about "Jewish" and "Aryan" mathematics. The core concepts reissued the antiformalist position of earlier times. The "positive, German type" would be intuitive, realistic, close to *Volk* and life; while the "countertype" would be a highly sophisticated juggler of formal concepts with no bonds to reality and no other motive than his own success. A key concept was *Anschauung,* indicating a variety of associations from Kantian philosophy via mental imagery down to the work with concrete geometrical figures. *Anschauung* had an imprecise but fairly well established meaning in mathematics, mainly as geometrical intuition in drawing conclusions from mental imagery in geometry or analysis. As a heuristic tool *Anschauung* had come into disrepute by a series of spectacular findings during the second half of the nineteenth century that showed that such conclusions frequently were invalid. For example, the intuitive concept of dimension (such as, a line has dimension one, a surface dimension two, a solid dimension three) was put in question by the construction of a space-filling curve. *Anschauung* had since been pushed back into the role of a secret heuristics with no place in formal mathematical discourse. By praising *Anschauung,* Bieberbach attacked the esoteric, apparently meaningless character of formal mathematics, and expressed the quest for the visibility of meaning and for bonding mathematics to reality. All this was framed such that it had obvious relations to parts of Nazi ideology. Bieberbach attempted, so to speak, a scientific counterrevolution with the help of the new powers. His attempt to dominate mathematics in Germany failed, however. The mathematicians' association managed to exclude him in early 1935. He started a journal, *Deutsche Mathematik,* to create his own, competing organization, but did not find the support he needed from the Nazi ministry of education. By 1937 Bieberbach and his group were an ideological residue without substantial influence in the system of mathematics. The external reasons for the failure will be discussed in the second part of this chapter.

In Bieberbach's counterrevolutionary enterprise there were three mistakes that led to his failure. First, he related mathematics via *Anschauung* to the Nazi worldview as he understood it. That is, he opted for one of the two possible relations between science and politics. He almost explicitly rejected what may be called the "instrumentality approach": scientific knowledge is a potential means of power; it can be turned into means of production, destruction, or control. Bieberbach presented mathematics instead as part of the officially projected order of things. We may call this the "metaphorical approach." *Anschauung* as a way to relate to the world is basically metaphor. In mathematics it becomes the term for a certain method but is filled with a wealth of connotations. It can be and has been scientifically productive through the power of association. The same word could have been used in general Nazi ideology. The Nazi world view is largely based on *Anschauung:* the structure of the world is immediately visible; political power and hierarchy are made aesthetic in parades and architecture; friend and foe can be recognized by their noses, and so on. The metaphor is productive also in social and political thought. This double productivity of a metaphor frequently serves to relate scientific and political thinking. The ideological fallacy (frequently intended) occurs in the transfer of deductions or social legitimacy from one area to the other with help of the metaphor. A more obvious example from biology is the concept 'instinct,' which was productive in Nazi Germany, both politically and scientifically, and served to mutually reinforce parts of Nazi ideology and biological ethology (cf. Kalikow, 1983).

Bieberbach's mistake was to put his stakes on this side. Nazi rule ultimately was much more interested in the instrumentality of the sciences and mathematics than in the correct worldview. Fundamental to this mistake is that Bieberbach did not simply fight for personal power but really wanted the Nazi revolution in mathematics to take place. Like other revolutionaries in the movement, he was instrumental in establishing the power of the new rulers but was denied any substantial access to that power.

Bieberbach's second mistake was to attack the cognitive phenomenon of "modern" mathematics, not noticing that this was the definition of a well-established social system. Indeed the definition of this social system of mathematics by its cognitive product is a means to defend the system and to veil power structures and social relations within the system. To oppose this with an alternative concept like 'German mathematics' in order to take over social power, when in fact the alternative system carried with it very little power, appears as a rather idealistic and helpless strategy. It is the strategy of the socially marginalized, of the losers of a basic social change, who believe that the glory they see in the old flag still bears some of the old power.

This type of critique may be applied to "alternatives" in science today as well. Sciences need to give room on the margins to feed creativity into their system of mental production. But the basic "paradigm" or self-definition of mainstream science rests on the safe basis of a well-established and flexible social structure that does not change flags easily.

The third mistake leads into the question of the relation of mathematical traditionalism (or intuitionism) and political nationalism (or fascism). Colleagues held that Bieberbach's belief in a "German" mathematics was honest. If this is true—and it seems to be—he projected values and wishes of a general kind into a mathematical program. But mathematics is just one of many trades in which an individual must deal with the constraints of the system. Just as a plumber cannot satisfy all his emotional needs in his trade, neither can a mathematician hope to satisfy his need of connectedness of all life solely in his work. Secondly, the cognitive and social system of mathematics is structured such that these attempts are clearly rejected. Bieberbach implicitly attacked this structure and thus a central characteristic of the system. Unlike the student revolutionaries of the Nazi movement, calling indiscriminately for a "politicized science," he could have known better.

But whence the possibility of such a projection, and why did he not know better? Bieberbach observed, historically and correctly, the existence of two "styles" in mathematics, formalist and intuitionist (*anschaulich*), and the present dominance of the former. Bieberbach, along with most mathematicians, associated this intuitionist "style" with the intuitionist program for the foundation of mathematics (developed by L. E. J. Brouwer mainly in the twenties) which attempted a reconstruction of what is meaningful in mathematics from the prime intuition of the one-after-the-other. Such a construction was highly restrictive for the existing mathematics and, therefore, never found a normative role (for which it was constructed). This program wanted to construct roads that would carry "meaning" into every remote branch of mathematics, the meaning that the relation to elementary mental action gave. The relation between technical intuitionism and the *anschaulich* style again was metaphorical. The metaphor normatively indicated that mathematics must not be cut off from the elements of human life. Brouwer's program and the older intuitionist style indicated that this was possible. And what could be more convincing to a mathematician than a mathematics that had shown its possibilities, plus a technically developed program of foundations that would grant validity to the results? The problem was that the social system of mathematics was (and is) interested not in meaning and motive but rather in production. The competing formalist style was obviously productive, and the corresponding foundational research program, Hilbert's "metamathematics," imposed no restrictions on existing mathematical theories except the demand for logical coherence. Hilbert attempted to build a wall against the intrusion of logical doubt by proving (with finite means) that no logical contradictions could occur. In this program a mathematical concept "existed" as soon as it was noncontradictory. This would grant a maximum of freedom in mathematical production. Although Hilbert's program in the strict sense failed, it was strong enough to reject any traditionalist normative claims to be pushed into mathematics via the critique of foundations.

This struggle was intense both because of the technical problem of avoiding logical contradictions like the paradoxes of set theory and the competing styles, as well as the fact that it incorporated positions in disciplinary policy. Hilbert justified his program with the "pre-established harmony between mathematics and nature." This can be translated into the harmony of the forces of the market. Formalist mathematics, it says, is productive and will sell. This accepts the social differentiation of mathematics and the social universality while demanding a maximum of autonomy for the "pure" core. It is obviously compatible with disciplinary internationalism and nonpolitical specialism.

The intuitionists, on the other hand, saw fragmentation instead of differentiation, decadence instead of productive freedom, esotericism and discord instead of the harmony. The policy of the discipline should care for inner unity, for meaning that would shine into teaching and application, for showing utility and cultural value of mathematics for humanity and for the nation. A mathematics that would be inherently tied to the essentials of human life would be the best means to support such a political program.

Through this connection there is a clear potential to relate mathematical style and a foundational program to political ends. Constructivism, and the metaphor of "action" involved in it, gives an immediate relation to political or moral thought. The connecting hypothesis is that somebody who wants to relate his life and work as a mathematician to his political or moral views is looking for a highly connected and pervasive system of interpretation of facts, values, and aims. He would rather opt for some sort of constructivism or intuitionism than for formalism, which defies the connectedness of a worldview by walling off mathematics from the real world. Formalism seems to fit with liberal, pragmatic, or politically uninterested positions. Intuitionism, on the other hand, seeks to connect mathematics and the real world of human action, allowing for an integration of professional political views and values. Such a connection, formalism-liberalism and intuitionism-reactionary political romanticism, is obviously by no means necessary or inevitable. Counterexamples are easily found. But it is a potential that was to a significant extent realized at that time. In general, the social system and the cognitive structure of mathematics are opposed to such connections, defending the purely professional identity of the system (not of the individuals). Furthermore, the metaphorical connection of politics and mathematics, with its basis in "action," does not prescribe specific details of either side, politics or science. Thus intuitionism need not be prefascist.

What is left to be explained historically is why this potential was realized at that point of history. Very briefly said, it is the experience of the losses through modernization that was complemented and reinforced by the experience of the German breakdown in World War I and the political, social, economic, and cultural crisis of the following years. The feeling of "crisis" was pervasive, and what was then called the "foundational crisis" of mathematics was part of it. In the general situation a new understanding of the world had to be found, or the old one had to be reconstructed and defended against the power of modernization. Nationalism, an emotional crutch in Germany in face of the never completely realized national unity, was one of the means to preserve and reerect a political identity after the experience of the war, the revolution, and the Weimar crises. Many mathematicians were nationalistic, some let this show in their professional work, and a few made it explicitly part of their vision of mathematics and its social status. The longing for unity and identity was transported into the mathematical profession by pressing for unity there and for bonds to the real world and the political task of national resurrection. And this was leading to fascism, to the violent attempt to construct a social world where it was easy to feel at home and safe, where there was unity and connectedness, a visible and simple structure of life and society, where it was easy to see who and what was responsible for the good and the bad.

But just this longing, this view of the world of mathematics, was marginalized by the force of its social and cognitive modernization. Mathematicians on this margin, not Bieberbach alone, took the chance for a revolt presented by the Nazi victory of 1933, and they were willing to use violence, more or less blinding themselves or being ruthlessly opportunistic. This explanation is, I believe, in a very similar way applicable to "German" physics (Beyerchen, 1977; Richter, 1980), which was more prominent because of the stronger cultural meaning of the mechanistic worldview and its revolution through quantum and relativity theory.

ADAPTATION AND RESISTANCE OF MATHEMATICS
IN NAZI GERMANY

The analysis of the process of integration of mathematics into the society of Nazi Germany will focus on the three professional societies. The story has been told in some detail elsewhere (Mehrtens, 1985a). I shall concentrate now on the basic interpretations. The most important organization was the German mathematicians' association, the *Deutsche Mathematiker-Vereinigung* (DMV). Bieberbach was the permanent secretary of the DMV and one of the editors of its journal. The social status of the discipline was in danger because of the anti-intellectual thrust of Nazi ideology, and its autonomy was endangered by the quest for pervasive politicking. Thus one would expect a combination of adaptive and defensive moves. Adaptation was necessary for the social system since it depends wholly on the material resources granted by the society in which it exists. But professional autonomy and international relations also had to be preserved as essential constituents of the system. In Nazi Germany conflict could not be evaded. The most interesting conflict is that with Bieberbach. He represented the danger of a loss of autonomy and, with his antimodern aims, a loss of productivity. In some individual cases the fight against Bieberbach was or might have been motivated by morals and politics. The social system, however, knows nothing of morals and politics, and its representatives in general act accordingly.

In early 1934 Bieberbach had printed in the DMV journal an open letter that he had written as an answer to a sharp attack against his ideas of "German" mathematics by a Danish mathematician. He did this without the consent of his coeditors and without the knowledge of the president of the DMV. Since Bieberbach could not be voted out of his office, this event appeared to offer a way to force him to resign. When the matter was discussed at the 1934 business meeting, however, the DMV was not successful in separating this question of professional behavior from matters of general politics. The resolution that was passed "condemned sharply" the attack against Bieberbach "in as far as one can find in it an attack at the new German state and National Socialism" and it merely "regretted" Bieberbach's behavior (my translation).

In the atmosphere of fear created by the political and racial purges, the DMV adapted to the pressure of politicking from the "new German state." Internally, where cognitive autonomy was endangered, the defense was stronger. The move to make Bieberbach führer of the society was voted down by a large majority. And an amendment of the statute was passed, and while the amendment looked somewhat like the national socialistic principle of hierarchical leadership (*Führerprinzip*) it was meant to eliminate Bieberbach by giving the president the right to dismiss officers of the association. The need to file the amendment at a special court, for which Bieberbach as secretary was responsible, led to a struggle, in which Bieberbach finally tried to use the power of the ministry and the help of some Nazi colleagues. But the result was that both the president of the DMV and Bieberbach were forced to resign. A politically reliable new president, George Hamel, came into office, and the DMV had to declare its loyalty to the state without being forced to amend the statutes.

Obviously, the ministry was not interested in making Bieberbach führer of the German mathematicians. To the contrary, the political powers were willing to accept the professional autonomy of the specialists, as long as they were sufficiently loyal to the state. Thus the mathematics profession, by the threat coming from the Nazi revolutionary Bieberbach, was pushed into a compromise with the ministry, which was equally national socialist but apparently less radical. This type of development can be found in many places. The social historian T. Mason expressed it thus: "Conservative forces preserved social and state order for National Socialism by saving it initially from National Socialism" (Mason, 1978:106 my translation). If Bieberbach had succeeded, the productivity of German mathematics would have been destroyed. Thus the DMV preserved a functioning mathematics for the Nazi state by protecting it from a Nazi revolution of mathematics. The immediate motive, which was most prominent in the fight against Bieberbach, was the fear that the DMV would lose its large foreign membership and thus its status in the international world of mathematics. This motive of internationalism embraced the preservation of productivity (as enforced by international communication), of autonomy (as enforced by recourse to structures out of reach of national powers), and of privileges as a resource of the disciplinary exchange system. We shall find such motives even clearer in the society of applied mathematicians.

A purely national organization was the Reich-association of mathematical societies—briefly called *Reichsverband* (MR). The MR was closely connected with the DMV and acted as an autonomous department for mathematical interest policies. The object of its activities was teacher education and school curricula. Since its foundation in 1921, close relations were also maintained with the association for the advancement of the teaching of mathematics and natural sciences, the so-called *Forderverein*. The MR was based in Berlin and its president was Georg Hamel, mathematics professor at the *Technische Hochschule*, Berlin.

Mathematics drew its social legitimacy mainly from its place in secondary education. For this reason the MR had been founded, and for this reason the MR followed the *Forderverein* in greeting national socialism enthusiastically in 1933. Both organizations bowed to Nazi ideology and hastened to declare that mathematics and natural sciences were rooted in the German soul and thus were an indispensible part of national socialist education. Even if Hamel was a nationalist, there is little reason to assume that he believed in what he said. He appears as the salesman of his discipline. The status of mathematics in the curricula of secondary schools appeared to be in danger. To defend that status while remaining allied with the teachers, the MR followed the submissive policy of the *Forderverein*. Furthermore, the members of the MR council were not likely to choose the way of political opposition. Indeed the activities of the MR and of Hamel met no official opposition from the DMV, other than Bieberbach. This kind of salesmanship did not endanger professional autonomy or productivity. Furthermore it was institutionally dissociated from the DMV, so there appeared to be little need to feel responsible.

One of the first tasks the MR set for itself was the production of a handbook with exercises that would show the value of mathematics to the new powers. The book included military, economic, and ideological topics of all kinds, for example: "It requires six million reichsmark to build a lunatic asylum; how many new homes, each costing fifteen thousand reichsmark , could have been built for this cost?"

What was shown here and in numerous similar books was the instrumentality of mathematics. On one hand, mathematics in school could be an instrument of transport for Nazi ideology. On the other hand, mathematical, calculating rationality obviously could be used to gain power over the social and natural world in the fascist manner. The bureaucratic mentality of the administrators of slave labor and of death camps was prefigured in such exercises. The "scientific" equivalent can be found in social statistics, for example, in the work and career of Siegfried Koller (cf. Aly and Roth, 1984).

The problem of analysis in this case, which needs further work, lies in the moral and political meaning of scientific instrumentality and in the social organization of social legitimation. The place of the MR on the margin between school and university mathematics was necessary for its political function. Being subject to pressures from the teachers, who out of their specific social and professional situation were tending strongly toward national socialism, the MR forced the integration of mathematics into the Nazi system. Ideological sales-manship of the first years and the concentration on applied mathe-matics that followed were tolerable for the system of scientific pro-duction in mathematics. There was no systemic resistance against this movement of political adaptation. One could argue that, to the con-trary, social universality and differentiation of modern mathematics was a precondition for such adaptation (compare the argument at the end of this article).

The change around 1936-37 towards preference for the loyal and competent specialist, reinforced by the serious shortage of technical personnel, found complementary developments in the MR and the DMV. Both associations cooperated in creating the image of what was called the "industrial mathematician" and developed a new univer-sity examination and curriculum that became effective in 1942 but had been discussed throughout the second half of the thirties. This move was adaptive to the chances offered by the need for technical specialists and by the fact that a few mathematicians were indeed working in industry, having been pushed into such careers by politi-cal pressures inhibiting academic careers. Basic autonomy was pre-served, since the offer was to educate competent specialists. The mathematicians in universities and institutes of technology had to change their teaching programs but hardly their fields of research. And what was taught and had to be put in textbooks was basic ap-plied mathematics, thus not subject to the competence of engineers or physicists. In fact, the changes in the teaching were, at least in univer-sities, slight. Even the thrust for the "industrial mathematician" and applied mathematics appears to have been more propaganda than re-ality. Involved were also institutional conflicts between the represen-tatives of various research specialties on one hand (that is, more or less applied) and between mathematicians at institutes of technology and at universities on the other. It has been noted that a change from the "indoctrination function" to the "qualification function" took place in school mathematics (Nyssen, 1979).

For applied mathematicians the situation was in general differ-ent. Their society was the GAMM, the society for applied mathemat-ics and mechanics. Here no problems like those Bieberbach created for the DMV occurred. But the GAMM had to face the racial purges. When in 1933 two of the three leading officers of the society decided to resign because of their Jewish descent, the president Ludwig Prandtl asked his colleague Erich Trefftz whether he would take over the presidency. Trefftz however urged Prandtl to keep his position, adding: "If we are forced to exclude Jewish members, I would hold the dissolution of the society to be the most honorable reaction." Prandtl answered at length, arguing that the GAMM gave mechanics the adequate place besides mathematicians and physicists. This need he wrote, "persists today more than ever." "With considerations of honour this, to my feeling, has nothing to do since it concerns simply a necessity of the discipline" (cf. Mehrtens, 1985:96ff.). After the war Prandtl called himself an "unpolitical German." In this case he acted as the "unpolitical" representative of his discipline. The "necessity of the discipline" was its survival in face of competition from other fields and in face of the growing demand for aircraft research. The moral position of Trefftz was brushed aside and did not reoccur. Prandtl, his institute, and his discipline had quite a career in the following years.

Prandtl actively defended the Jewish membership of the society. The main motive, however, lay not in human or moral values but in the attempt to ensure and enlarge the international standing of the society and the discipline. The plan for an international congress for mechanics in Germany could be sold to the aircraft ministry as a necessity for productive aircraft research. The exclusion of Jews would, Prandtl argued, jeopardize the congress. In the end, these plans failed, because the ministry of education was not willing to unconditionally admit Jews to the congress. Jewish members of the GAMM were excluded by 1938, formally on the basis of unpaid dues.

The GAMM was forced to give up much of its international relations. The combined disciplines of mechanics and applied mathematics, however, gained resources and status from the demand for their products and their competence. There were fairly few problems of integration into the Nazi system. Scientific internationalism appears as a professional value motivating resistance against political adaptation but it fell victim to the racist and imperialist politics of Nazi Germany.

The problem of analysis in this case has been merely touched upon. Viewed from the system of mathematics, the mechanics Prandtl represented is a different and competing discipline. But applied mathematics had neither an autonomous social organization nor the cognitive core for a clear disciplinary identity. In the course of the foundation of the GAMM, Richard von Mises, the applied mathematician, formulated the identity of the field in terms of "practical needs" (von Mises, 1921). Applied mathematics thus is much more dependent on the demand for its products. The lack of cognitive coherence has to be compensated by stronger integration into its market environment. The combination with mechanics and thus with aircraft research turned out as the best option during the thirties. Like the MR, the GAMM was an institution on the margin of the social system of mathematics, mediating influences and interests in both directions.

For a fairly complete survey of the social system of mathematics in Nazi Germany, the situation during the war should be analyzed. The empirical basis is, as yet, too small. It should just be noted here that matters of mathematics education lost prominence while application and applicability dominated disciplinary politics. Such politics were possible and promising after 1942, when the time of quick victories was over. Structures and functions of the disciplinary system remained basically unchanged but were obviously conditioned by the war. The new possibilities and the poor general organization of military research and development led to more and sharper conflicts within the discipline. Different groups and institutions cooperated and competed with wholly selfish interests, bound together mainly by the common fear of the end. The president of the DMV who was, after a change in the statutes, reelected in 1937 became head of the newly founded mathematics branch of the *Reich* research council and managed to raise considerable funds for applied basic research. He was involved in calling mathematicians back into the institutes from military service and finally even succeeded in creating a national research institute for mathematics that became a refuge for many mathematicians during the last months of the war and later. This success was possible through close cooperation with many Nazi officials, including involvement in the creation of a mathematical institute in a concentration camp where scientific slave labor was exploited.

CONCLUSION

Mathematics was integrated into the social and political system of Nazi Germany. Specificity was the integrative mode for this society that was disintegrating and in which terror, violence, and fear were the most important social bonds. I have hardly touched upon the problems of individual and social psychology involved in the understanding of this period. How, for example, were mathematicians able to blind themselves to the suffering of their Jewish friends and colleagues? This is an important question since the social system of mathematics was able to survive with the loss of some 20 to 30 percent of its members as long as this remained tolerable to the remaining mathematicians. In fact, there are only very few cases of mathematicians who emigrated or isolated themselves because of this experience. From the point of view of the social system of mathematics, this is a question of environment relations. The mathematician as an individual (as a psychic system) is part of its environment. This dissociation of the individual scientist from the system of the discipline is certainly an unsolved theoretical problem in the sociology of science, but it appears as a very useful move for the (theoretically eclectic) historian to solve some of the historiographical problems involved in the present subject.

Rather, the social system in which individuals play their various roles is the perspective of this chapter. Fear, opportunism, and partial identification with Nazi politics have certainly been integrative forces on the mathematics system acting through the individuals. Further there was the move of the radical attack from revolutionary students and from Bieberbach pressing the system to seek cooperation with those blocks of power in which bureaucratic or technocratic rationality persisted (for example, in ministerial bureaucracy or the air ministry). For such movements the social systems of the discipline preferred individuals who would manage the balance between adaptation and defense of the system's identity, as in the Prandtl-Trefftz case. There also was the offer of intensified technical development, to which mathematics reacted by enlarging its applied subsystem and by restructuring its educational functions.

Looking for the conditions of adaptability we find in the three professional associations a functional variance. The MR was largely in charge of the educational legitimation of the discipline. It fulfilled its task by deeply bowing to the new powers and following the main trends in the associated teaching profession. This submission to the political powers did not jeopardize any central necessities of survival of the discipline, since it had little to do with the system of knowledge production and was a purely national organization. The MR's declaration of loyalty, visibly coming from the closest vicinity of the DMV, relieved the latter from political pressure and gave the chance to defend the discipline against the radical Bieberbach. The fact that Hamel, the head of the MR, became the compromise president of the DMV after the resignation of Bieberbach and the former president is an expression of this relation.

The DMV represented all of mathematics, centered around its pure core. It declared its loyalty in a fairly strong way when it reached the compromise with the ministry. That meant some politicking, since afterwards disciplinary identity and the borderlines of the system were less rigid. At the same time new possibilities of legitimation by utility opened up. The DMV, together with the MR, concentrated on the educational task, that is, on the production of mathematical competence to be applied in other social systems, thus preserving autonomy and legitimacy of pure mathematics as the indispensible basis for such competence. Since this went beyond the traditional task of the MR, the DMV took over much of the disciplinary policy involved, even more when the president was made reelectable in order to be politically efficient. And the then permanent president up to the end of the war worked in the interest of the survival of mathematics as a pure but applicable discipline.

The production of applied mathematics, that is, of knowledge oriented towards specific tasks outside mathematics, was largely represented in the GAMM. This work could be easily identified with immediately visible development of technology in aircraft production, a technology that was an obvious means and expression of power. Thus there was a specific relation to Nazi ideology that made things easy for the GAMM. In its defense of international communication, the GAMM not only fought for its institutional status and its productivity but also presented visibly the necessities of scientific production for all of mathematics to the political powers.

Finally the Bieberbach group in Berlin has to be seen within this view of functional integration. Mathematicians, with the help of the ministry, managed to deprive Bieberbach of any substantial influence. But they were not interested in eliminating him completely; rather there were certain fields of cooperation. Large parts of traditional Nazi ideology became residual and had little to do with the social and economic realities of the country. They persisted however in providing an integrating function for the Nazi-movement and its claim to complete power. The Bieberbach group related mathematics to that element of the Nazi system and kept for some time its role as an ideological showcase and playground. It lost that role only late in the war.

The variations of status and functions within the system of mathematics corresponded to variations in the definition of cognitive identity. To describe this in any detail again is an open problem. It is obvious, however, that the core function of this identity, namely, to handle the difference to the environment and to preserve and, if necessary, adapt borderlines, remained with pure mathematics. The toleration of the Bieberbach group appears as such a temporary adaptation of the borderline. The playground for the Bieberbach type of deviance was necessary in this society. It remained, however, potentially dangerous to modern professional identity. This is part of the explanation for that fact that after the war Bieberbach was the only productive mathematician who never found a university position again. He became the symbol for Nazi deviance in mathematics, while the disciplinary politicians of the field, despite their intense collaboration with the Nazi system, remained honored representatives of German mathematics.

Finally I would like to address the question of the political meaning of this account. The Nazi experience has been used frequently to argue that natural science and mathematics are inherently democratic. I am afraid this is not the case. There is a tendency towards political liberalism inherent in the double function of international communication as being a condition of productivity as well as of autonomy in face of national powers. But the sciences are dependent on national societies for their material existence. They will adapt to political and social changes as long as there is the chance to preserve existence. The irrational, regressive side of German fascism was an existential danger to mathematics. Its technocratic, imperialist side, however, was compatible enough with the survival of the discipline. Mathematics offered a means of technological advancement, for instance, by producing the aerodynamical theory and the calculational means for the V2 rocket, and it could offer means of social control (for example, in the statistics of inheritance). And mathematicians did offer these services without becoming outcasts of the discipline. To the contrary, they also found a market with the victorious powers, immediately after the war.

Cognitive and social universality together with social differentiation of the system were the basis for survival of the discipline under extreme political and social changes. Universality makes the products saleable to any political power interested in new means of power. Social differentiation takes care of adaptive functions and at the same time ensures that individual moral or political aims will not jeopardize the existence of the system. As long as it is possible, the disciplinary system exchanges its products, loyalty, and political neutrality for material resources and legitimacy from its environment. I cannot find any reason why mathematics, and any other science, should not find a perfect partner in technocratic fascism. Except, perhaps, because of the need of such regimes to use means of social control that tend to destabilize social subsystems. It is this observation that makes the subject of the second part of this chapter more interesting and more important as a subject of further research in the history and sociology of science than the analysis of the obvious deviance in Bieberbach's "German" mathematics.

*This chapter is a result of a research project funded by the Stiftung Volkswagenwerk. All translations from the German are mine. Due to the character of the chapter as a theoretical survey, historical statements are not fully annotated. The reader is referred to the references.

NOTE

This chapter was first published in *Sociological Inquiry* 57, 2:159-82. Permission for republication granted by Texas University Press is gratefully acknowledged.

BIBLIOGRAPHICAL NOTE, 1990

The body of studies on natural sciences and mathematics has substantially grown during the last five years. On interpretations of the history and structure of the Third Reich in general see Kershaw, 1985; for the structure of political power see Hirschfeld and Kettenacker, 1981. For universities and natural sciences under National Socialism see, for example, Becker, Dahms, and Wegeler, 1987; Lundgreen, 1985; Muller-Hill, 1987, 1988; Seier, 1984, 1988; and Walker, 1989a, 1989b. The surveys I have presented (Mehrtens 1979, 1980) are still largely valid and most open questions indicated there are still open. Mathematics and national socialism has been studied by Segal, Siegmund-Schultze, and me. For Ludwig Bieberbach, compare Lindner, 1980; and Mehrtens, 1987; his conflict with Hilbert in 1928 is described by van Dalen, 1990. For applied mathematics and applications see Mehrtens, 1986; for school mathematics see Mehrtens, 1989; for the publication system and especially the abstracting journals see Siegmund-Schultze, 1986a; for the imperialist vision of German mathematicians in the early war years see Siegmund-Schultze, 1986b. On the emigration of mathematicians see Fletcher, 1986; Reingold 1981; and Rider, 1984. The systems theoretical approach to mathematics is also developed in Mehrtens, 1988; and by MaaB, 1988. Stichweh (1984) has in an impressive study applied this view to nineteenth-century physics. The programmatic sketch given in this article of the structure and role of modernism and traditionalism in mathematics has meanwhile developed into a book (Mehrtens, 1990).

REFERENCES

Aly, G. and K. H. Roth
   1984 *Die restlose Erfassung: Volkszählen, Identifizieren, Auusondern im Nationalsozialismus.* Berlin: Rotbuch.

Becker, H., H.-J. Dahms, and C. Wegeler
   1987 *Die Universität Göttingen unter dem Nationalsozialismus: Das verdrangte Kapitel ihrer 250 jährigen Geschichte.* Munich: K. G. Saur.

Beyerchen, A.
1977 *Scientists Under Hitler: Politics and the Physics Community in the Third Reich.* New Haven: Yale University Press; German: *Wissenschaftler unter Hitler: Physiker im Dritten Reich.* Frankfurt: Ullstein 1982.

Bourdieu, P.
1975 "The Specifity of the Scientific Field and the Social Conditions of the Progress of Reason." *Social Science Information* 6:19-47.

Dalen, D. van
1990 "The War of the Frogs and the Mice, or the Crisis of the Mathematische Annalen." *The Mathematical Intelligencer 12, 4:17–31.*

Fletcher, C. R.
1986 "Refugee Mathematicians: A German Crisis and a British Response, 1933-1936." *Historia Mathematica* 13:13-27.

Geuter, U.
1984 *Die Professionalisierung der deutschen Psychologie im Nationalsozialismus.* Frankfurt: Suhrkamp.

Heinemann, M., ed.
1980 *Erziehung and Schulung im Dritten Reich.* Vol. 2: *Hoschschule, Erwachsenenbildung.* Stuttgart: Klett-Cotta.

Herf, J.
1984 *Reactionary Modernism: Technology, Culture, and Politics in Weimar and the Third Reich.* Cambridge: Cambridge University Press.

Hirschfeld, G., and L. Kettenacker, eds.
1981 *Der "Führerstaat": Mythos und Realität.* Stuttgart: Deutsche Verlags-Anstalt.

Kalikow, T. J.
1983 "Konrad Lorenz's Ethological Theory: Explanation and Ideology 1938-1943." *Journal for the History of Biology* 16:39-73.

Kershaw, I.
1985 *The Nazi Dictatorship: Problems and Perspectives of Interpretation.* London: Arnold. German: *Der NS-Staat: Geschichtsinterpretationen im Überblick.* Reinbek: Rowohlt, 1988.

Kline, M.
1980 *Mathematics: The Loss of Certainty.* New York: Oxford University Press.

Kuhn, T. S.
1971 *The Structure of Scientific Revolutions.* 2nd ed. Chicago: University of Chicago Press.

Lindner, H.
1980 "'Deutsche' und 'gegentypische' Mathematik: Zur Begründung einer

'arteigenen' Mathematik im Dritten Reich durch Ludwig Bieberbach." Pp. 88-15 in Mehrtens and Richter (1980).

Luhmann, N.
1984 *Soziale Systeme: GrundriB einer allgemeinen Theorie.* Frankfurt:
Suhrkamp.

Lundgreen, P., ed.
1985 *Wissenschaft im Dritten Reich.* Frankfurt: Suhrkamp.

MaaB, J.
1988 *Mathematik als Soziales System: Geschichte und Perspektiven der Mathematik aus systemtheoretischer Sicht.* Weinheim: Deutscher Studien Verlag.

Mason, T. W.
1978 *Sozialpolitik im Dritten Reich: Arbeiterklasse und Volksgemeinschaft.* 2nd
ed. Opladen: Westdeutscher Verlag.

Mehrtens, H.
1979 "Die Naturwissenschaften im Nationalsozialismus." Pp. 427-43 in
Rürup, Reinhard (ed.), *Wissenschaft und Gesellschaft: Beiträge zur
Geschichte der Technischen Universität Berlin 1879-1979.* Vol. 1. Berlin:
Springer.
1980 "Das 'Dritte Reich' in der Naturwissenschaftsgeschichte: Liter-
aturbericht und Problemskizze." Pp. 15-87 in Mehrtens and Richter
(1980).
1985 "Die 'Gleichschaltung' der mathematischen Gesellschaften im nationalsozialistischen Deutschland." *Jahrbuch Überblicke Mathematik*
83-103; English: "The 'Gleichschaltung' of Mathematical Societies in
Nazi Germany." *The Mathematical Intelligencer* 11,3(1989):48-60.
1986 "Angewandte Mathematik und Answendungen der Mathematik im
nationalsozialistischen Deutschland." *Geschichte und Gesellschaft*
12:317-47.
1987 "Ludwig Bieberbach and "Deutsche Mathematik.'" Pp. 195-241 in E.
Phillips (ed.), *Studies in the History of Mathematics.* MAA Studies in
Mathematics, vol. 26. Washington, D.C.: The Mathematical Association of America.
1988 "Das soziale System der Mathematik und seine politsche Umwelt.
"*Zentralblatt für Didaktik der Mathematik* 20,1:28-37.
1989 "Nationalsozialistisch eingekleidetes Rechnen: Mathematik als Wissenschaft und Schulfach im NS-Staat." Pp. 205-16 in R. Dithmar
(ed.), *Schule und Unterrichtsfächer im Dritten Reich.* Neuwied: Luchter-
hand.
1990 *Moderne—Sprache—Mathematik: Eine Geschichte des Streits um die
Grundlagen der Disziplin und des Subjekts formaler Systeme.* Frankfurt:
Suhrkamp.

Mehrtens, H., and S. Richter, eds.
1980 *Naturwissenschaft, Technik und NS-Ideologie; Beiträge zur Wissenschafts-geschichte des Dritten Reiches.* Frankfurt: Suhrkamp.

Mises, R. von
1921 "Über die Aufgaben und Ziele der angewandten Mathematik." *Zeitschrift für angewandte Mathematik und Mechanik* 1:1-15.

Muller-Hill, B.
1987 "Genetics after Auschwitz." *Holocaust and Genocide Studies* 2:3-20.
1988 *Murderous Science: Elimination by Scientific Selection of Jews, Gypsies, and Others, Germany 1933-1945.* Oxford: Oxford University Press.

Neumann, F.
1944 *Behemoth: The Structure and Practice of National Socialism.* New York: Oxford University Press.

Nyssen, E.
1969 *Schule im Nationalsozialismus.* Heidelberg: Quelle & Meyer.

Reingold, N.
1981 "Refugee Mathematicians in the United States of America, 1933-1941: Reception and Reaction." *Annals of Science* 38:313-38.

Richter, S.
1980 "Die 'Deutsche Physik.'" Pp. 116-41 in Mehrtens and Richter (1980).

Rider, R.
1984 "Alarm and Opportunity: Emigration of Mathematicians and Physi-cists to Britain and the United States, 1933-45." *Historical Studies of the Physical Sciences* 15:107-76.

Schappacher, N.
1987 "Das mathematische Institut der Universität Göttingen 1929-1950." Pp. 345-73 in Becker, Dahms, and Wegeler (1987).

Schwabe, K. ed.
1988 *Deutsche Hochschullehrer als Elite 1815-1945.* Boppard: Boldt.

Segal, S. L.
1986 "Mathematics and German Politics: The National Socialist Experi-ence." *Historia Mathematica* 13:118-35.

Seier, H.
1984 "Universität und hochschulpolitik im nationalsozialistischen Staat." Pp. 143-65 in K. Malettke (ed.), *Der Nationalsozialismus an der Macht.* Göttingen: Vandenhoek & Ruprecht.
1988 "Die Hochschullehrer im Dritten Reich." Pp. 247-95 in Schwabe (1988).

Siegmund-Schultze, R.

1984 "Theodor Vahlen—zum Schuldanteil eines deutschen Mathematikers am faschistischen Mißbrauch der Wissenschaft." *NTM Schriftenreihe für die Geschichte der Naturwissenschaften, Technik und Medizin* 21,1:17-32.

1986a "Beiträge zur Analyse der Entwicklungsbedingungen der Mathematik im faschistischen Deutschland unter besonderer Berucksichtigung des Referatewesens." Humbold-Universität Berlin. Dissertation.

1986b "Faschistische Pläne zur 'Neuordnung' der europäischen Wissenschaft: Das Beispiel Mathematik." *NTM Schriftenreihe für die Geschichte der Naturwissenschaften, Technik und Medizin* 23,2:1-17.

Stichweh, R.

1984 *Zur Entstehung des modernen Systems wissenschaftlicher Disziplinen: Physik in Deutschland 1740-1890*. Frankfurt: Suhrkamp.

Walker, M.

1989a *German National Socialism and the Quest for Nuclear Power*. Cambridge: Cambridge University Press; German: *Die Uranmaschine: Mythos und Wirklichkeit*. Berlin: Siedler 1990.

1989b "National Socialism and German Physics." *Journal of Contemporary History* 24:63-89.

13

The Social Life of
Mathematics

INTRODUCTION

Mathematicians and philosophers of mathematics have long claimed exclusive jurisdiction over inquiries into the nature of mathematical knowledge. Their inquiries have been based on the following sorts of assumptions: that Platonic and Pythagorean conceptions of mathematics are valid, intelligible, and useful; that mathematical statements transcend the flux of history; that mathematics is a creation of pure thought; and that the secret of mathematical power lies in the formal relations among symbols. The language used to talk about the nature of mathematical knowledge has traditionally been the language of mathematics itself; when other languages (for example, philosophy and logic) have been used, they have been languages highly dependent on or derived from mathematics. By contrast, social talk takes priority over technical mathematical talk when we consider mathematics in sociological terms.

Sociological thinking about mathematics has developed inside and outside of the mathematical community. In the social science community, it is manifested in insider, professional sociology. In the mathematical community, the everyday folk sociology of mathematicians became better articulated as mathematical work became better organized and institutional continuity was established beginning in seventeenth-century Europe. Eventually, some mathematicians who were especially conscious of the social life of the mathematical community began to write social and even sociological histories of their field (Dirk Struik, for example) or exhibit that consciousness in their mathematical programs (as in the cases of the constructivists and the group of mathematicians known as N. Bourbaki). It is therefore no longer obvious that technical talk can provide a complete understanding of mathematics.

The decline of Platonic, Pythagorean, formalist, and foundationalist prejudices has opened the door for social talk about mathematics. But the implications and potential of social talk about mathematics have yet to be realized. My task in this chapter is programmatic: it is to sketch the implications and potential of thinking and talking about mathematics in sociological terms.

There is a sociological imperative surfacing across a wide range of fields that is changing the way we view ourselves, our world, and mathematical and scientific knowledge. This is not disciplinary imperialism; the sociological imperative is not the same as the discipline or profession of sociology. It is a way of looking at the world that is developing in the context of the modern experience, a mode of thought emerging out of modern social practice. The basis for this Copernican social science revolution lies in three interrelated insights: *all* talk is *social*; the person is a *social structure*; and the intellect (mind, consciousness, cognitive apparatus) is a *social structure*. These insights are the foundation of a radical sociology of mathematics.

INTELLECTUAL ORIGINS OF THE SOCIOLOGICAL IMPERATIVE

The program for a radical sociology of mathematics, in which all talk about mathematics is social talk, begins with Marx's formulation of the insight that science is a social activity. In order to underline the theme of this chapter, I take the liberty of substituting the term mathematical for scientific in the following quotation:

> Even when I carry out [mathematical] work, etc., an activity which I can seldom conduct in direct association with other men—I perform a social, because human, act. It is not only the material of my activity—like the language itself which the thinker uses—which is given to me as a social product. My own existence is a social activity. (Marx, in Bottomore and Rubel, 1956:77)

This fundamental statement of the sociological imperative achieves its classical form in the closing pages of Emile Durkheim's *The Elementary Forms of the Religious Life*. Here, Durkheim initiates the transformation of the apparently obvious observations that the mathematician is a social being and that even his/her language is social into a nonobvious sociology of concepts.

Durkheim argues that individualized thoughts can be understood and explained only by attaching them to the social conditions they depend on. Thus, ideas become communicable concepts only when and to the extent that they can be and are *shared* (Durkheim, 1961:485). The laws of thought and logic that George Boole searched for in pure cognitive processes (see the discussion below) are in fact to be found in social life. The apparently purest concepts, logical concepts, take on the appearance of objective and impersonal concepts only to the extent that and by virtue of the fact that they are *communicable* and *communicated*—that is, only insofar as they are *collective representations*. All concepts, then, are collective representations and collective elaborations because they are conceived, developed, sustained, and changed through social *work* in social contexts. In fact, *all* contexts of human thought and action are social. The next intellectual step is to recognize that 'work', 'context', 'thought', and 'action' are inseparable; concepts, then, are not *merely* social products, they are *constitutively* social. This line of thinking leads to the radical conclusion that it is social worlds or communities that think, not individuals. Communities as such do not *literally* think in some superorganic sense. Rather, individuals are *vehicles* for expressing the thoughts of communities or "thought collectives." Or, to put it another way, *minds are social structures* (Gumplowicz, 1905:268; Fleck, 1979:39).

It is hopeless to suppose that social talk insights could be arrived at or appreciated by people immersed in technical talk. In order to understand and appreciate such insights, one has to enter mathematics as a completely social world rather than a world of forms, signs, symbols, imagination, intuition, and reasoning. This first step awakens us to "math worlds," networks of human beings communicating in arenas of conflict and cooperation, domination and subordination. Here we begin to experience mathematics as social practice and to identify its connections to and interdependence with other social practices. Entering math worlds *ethnographically* reveals the continuity between the social networks of mathematics and the social networks of society as a whole. And it reveals the analogy between cultural production in mathematics and cultural production in all other social activities (Collins, 1985:165).

The second step in this sequence occurs when we recognize that social talk and technical talk seem to be going on simultaneously and interchangeably. The third step brings technical talk into focus in terms of the natural history, ethnography, and social history of signs, symbols, vehicles of meaning, and imagination. The more we participate in the math worlds in which mathematicians "look, name, listen, and make," the more we find ourselves despiritualizing technical talk (Geertz, 1983:94-120). The final step in comprehending the sociological imperative occurs when we realize at last and at least in principle that technical talk *is* social talk.

Mathematical knowledge is not simply a "parade of syntactic variations," a set of "structural transformations," or "concatenations of pure form." The more we immerse ourselves ethnographically in math worlds, the more we are impressed by the universality of the sociological imperative. Mathematical forms or objects increasingly come to be seen as sensibilities, collective formations, and worldviews. The foundations of mathematics are not located in logic or systems of axioms but rather in social life. Mathematical forms or objects *embody* math worlds. They contain the social history of their construction. They are produced *in* and *by* math worlds. It is, in the end, math worlds, not individual mathematicians, that manufacture mathematics (cf. Becker, 1982).

Our liberation from transcendental, supernatural, and idealist visions and forces begins when the sociological imperative captures religion from the theologists and believers and unmasks it; it becomes final (for this stage of human history) when that same imperative takes mind and intellect out of the hands of the philosophers and psychologists. It is in this context of inquiry that the larger agenda of the sociology of mathematics becomes apparent. Durkheim set this agenda when he linked his sociological study of religion to a program for the sociological study of logical concepts. The first full expression of this agenda occurs in Oswald Spengler's *The Decline of the West*.

SPENGLER ON NUMBERS AND CULTURE

In one of the earliest announcements of a "new sociology of science," David Bloor mentions Oswald Spengler as one of the few writers who challenges the self-evident "fact" that mathematics is universal and invariant (Bloor, 1976:95). But Bloor says little more about Spengler's "Numbers and Culture" chapter in *The Decline of the West* than that it is "lengthy and fascinating, if sometimes obscure." Length and obscurity are apparently two of the reasons Spengler has been ignored as a seminal contributor to our understanding of mathematics. He is also considered too conservative, and even a fascist sympathizer and ideologue, by some intellectuals and scholars and thus unworthy of serious consideration as a thinker. But Spengler was not a fascist and certainly no more conservative or nationalistic than other scholars and intellectuals who have earned widespread respect in the research community (Max Weber, for example). And the interesting affinities between Spengler and Wittgenstein—and in fact Spengler's influence on this central figure in the pantheon of modern philosophy—are only now beginning to come to light. Of special interest in this respect is their common and widely overlooked ethical agenda (brought to my attention by the historian Peter John). But it is their common vision of an anthropology of mathematics that is of immediate interest (on Spengler, Wittgenstein, and other contributors to the anthropology of mathematics, see Restivo, 1983:161-75; Crumb, 1989).

There can be little doubt that part of the reason for the resistance to Spengler is that unlike other writers who have challenged the central values of Western culture (including Wittgenstein), Spengler is harder to address as a reasonable and recognizable opponent or ally. Contradictions and paradoxes in his work aside, he does not really want to play the games of modern science, culture, and philosophy. Those who do want to play these games cannot really use Spengler, even when they somehow can appreciate him. Bloor is a case in point. Whatever his admiration for and indebtedness to Spengler, in the end Bloor's sociology of mathematics is grounded in a defense of modern (Western) science and modern (Western) culture (Restivo, 1988:208).

Spengler's analysis does not assign a privileged status to Western culture or Western science. It is also important to recognize the significance of the priority he assigns to numbers. The first substantive chapter in volume 1, chapter 2, is on numbers and culture, and it identifies mathematics as a key focus of Spengler's analysis of culture. Since I have discussed Spengler's views at length elsewhere, I will be very brief in identifying the central tenets of his theory (Restivo, 1983:211-31).

Number, according to Spengler, is "the symbol of causal necessity." It is the sign of a completed (mechanical) demarcation; and, with God and naming, it is a resource for exercising the will to "power over the world." Because Spengler conceives of cultures as incommensurable (although he allows for the progressive transformation of one culture into another), he argues that mathematical events and accomplishments should not be viewed as stages in the development of a universal, world "Mathematics." A certain type of mathematical thought is associated with each culture—Indian, Chinese, Babylonian-Egyptian, Arabian-Islamic, Greek (classical), and Western. The two major cultures in Spengler's scheme, classical and Western, are associated, respectively, with number as magnitude (as the essence of visible, tangible units) and number as relations (with function as the nexus of relations; the abstract validity of this sort of number is self-contained; Spengler, 1926:chap. 2).

Spengler's theory of mathematics yields a weak and a strong sociology of mathematics. The weak form is the one that students of math worlds who have accepted the validity and utility of social talk about mathematics find more or less reasonable. It simply draws attention to the variety of mathematical traditions across and within cultures. The strong form implies the sociological imperative—the idea that mathematical objects are constitutively social.

THE WEAK SOCIOLOGY OF MATHEMATICS

The weak form of Spengler's theory is illustrated by the alternative mathematics discussed by Wittgenstein and Bloor (these are not, in fact, *alternatives* to modern mathematics but rather *culturally* distinct forms of mathematics) and by specific mathematical traditions (European, sub-Saharan, Chinese, and so on). The study of these traditions produces stories about the "mathematics of survival," sociocultural bases for the rise and fall of mathematical communities, ethnomathematics, the social realities behind the myths of the Greek and Arabic-Islamic "miracles," and the organizational revolution in European mathematics and science from the seventeenth century on. The episodic history of Indian mathematics can thus be shown to be related to, among other factors, the fragmented decentralization of Indian culture and the caste system. "Golden" ages in ancient Greece, the Arabic-Islamic world between 700 and 1200, seventeenth-century Japan, T'ang China, and elsewhere can be causally linked to social and commercial revolutions. And the centripetal social forces that kept mercantile and intellectual activities under the control of the central bureaucracy in China can be shown to be among the causes of China's failure to undergo an autochthonous scientific revolution.

The differences among and within mathematical traditions do not necessarily signify incommensurability. They are compatible with the concept of the long-run development or evolution of a "universal"

or "world" mathematics. But the strong form of Spengler's theory—
that mathematics are reflections of and themselves worldviews—is
another story.

BASIC PRINCIPLES OF THE
STRONG SOCIOLOGY OF MATHEMATICS

There is no ready, intelligible exhibit of the strong form of Spengler's theory of numbers and culture, that is, an example that would persuade a majority of mathematicians and students of math studies that indeed it is possible to describe and explain the content of the "exact sciences" in sociological terms. But there *are* signposts on the road to such an exhibit. Some of these signposts are ingredients of the sociological imperative; some of them are the results of the still very slim body of research in the sociology and social history of mathematics. Based on these signposts, we can begin to anticipate telling a story about mathematics in terms of the strong form of Spengler's theory. I begin with a brief preface.

Numbers arise in part from the facts that (1) things and processes in our world are experienced and can be treated as discrete; and (2) discrete things and processes form or can be arranged in sets: "From this point of view numbers may be considered abstractable properties of sets rather than, as some philosophers have maintained, creations of the mind" (Hawkins, 1964:12-13). Number, then, and in the first place, refers to things and processes we experience in the world. In this sense, *reference* is demonstrative, nominative, or ostensive. But what about the second type of reference Hawkins (1964:18) calls "propositional" or "descriptive"? Here, too, reference is to a material world, but an at least second-order one. The central materialistic point about abstract or pure mathematics is expressed in the general notion that "whenever men develop sufficiently elaborate instruments, these tend to become objects of interest in their own right" (Hawkins, 1964:36).

Can we then make sense in sociological and materialist terms of Hawkins's (1964:31) contention that mathematics is not *representative* but *presentative*, a "model" rather than a "description"? I am now ready to indicate how we might go about telling a sociological story about mathematics.

Mathematical workers use tools, machines, techniques, and skills to transform raw materials into finished products. They work in mathematical "knowledge factories" as small as individuals and as large as research centers and worldwide networks. But whether the factor is an individual or a center, it is always part of a larger network of human, material, and symbolic resources and interactions; it is always a social structure.

Mathematical workers produce mathematical objects, such as theorems, points, numerals, functions, and the integers. They work with two general classes of raw materials. One is the class of all things, events, and processes (excluding mathematical objects) in human experience that can be "mathematized". The second is the class of all mathematical objects. *Mathematical workers* work primarily or exclusively with raw materials of the first class. *Mathematicians* and especially pure mathematicians work primarily or exclusively with raw materials of the second class. The more specialized and organized mathematical work becomes, the greater the extent of overlap, interpenetration, and substitutability among mathematical objects, raw materials, machines, and tools. The longer the generational continuity in a specialized math world, the more abstract the products of mathematical work will become. It is also important to recognize the overlap between the class of mathematical objects, the class of raw materials for mathematical work, and the class of tools and machines for doing mathematical work.

As specialization increases and levels of abstraction increase, the material and social origins of mathematical work and products become increasingly obscure. In fact what happens is an intensified form of cultural growth. Cultural activity builds new symbolic layers on the material grounds of everyday life. The greater the level of cultural growth, the greater the distance between the material grounds of everyday life and the symbolic grounds of everyday life. Increasingly, people work on and respond to the higher symbolic levels. Imagine the case now in which a mathematical worker is freed from the necessity of hunting and gathering or shopping and paying taxes, and set to work on the purest, most refined mathematical objects produced by his/her predecessors. Under such conditions, the idea that pure mental activity (perhaps still aided by pencil and paper—already a great concession for the Platonist!) is the source of mathematical objects becomes increasingly prominent and plausible. Workers forget their history as creators in the social and material world and the history of their ancestors working with pebbles, ropes, tracts of land, altars, and wine barrels. Or, because they do not have the language for recording social facts, ignorance rather than failed memory fosters purist conclusions. The analogy with religious ideas, concepts, and thoughts is direct and was probably first recognized, at least implicitly, by

Durkheim. In Spengler, this idea achieves an explicit and profound expression.

Specialization, professionalization, and bureaucratization are aspects of the organizational and institutional history of modern mathematics. These processes occurred in earlier mathematical traditions but their scope, scale, and continuity in modern times are unparalleled. Their effect is to generate closure in math worlds. As closure increases, the boundaries separating math worlds from each other and from other social worlds thicken and become increasingly impenetrable. Specialized languages, symbols, and notations are some of the things that thicken the boundaries around math worlds.

Ultimately, in theory, if the process I have sketched goes on unchecked a completely closed system emerges. This is technically impossible. But as closure becomes more extreme, the math world (like any social world) becomes stagnant, then begins to deteriorate, and eventually disintegrates.

Some degree of closure facilitates innovation and progressive change; extreme closure inhibits them. The advantages of closure must therefore be balanced against the advantages of openness, that is, the exchange of information with other social worlds. Specifically, the danger for the social system of pure mathematics is that it will be cut off from the stimulus of external problems. If pure mathematicians have to rely entirely on their own cultural resources, their capacity for generating innovative, creative problems and solutions will progressively deteriorate. As a consequence, the results of pure mathematical work will become less and less applicable to problems in other social worlds.

The closure cycle reinforces the community's integrity; the opening cycle energizes it with inputs and challenges from other social worlds. The interaction of insider and outsider sociologies of mathematics, mathematical and nonmathematical ideas, and pure and applied mathematics are all aspects of the opening cycle that are necessary for creative, innovative changes in mathematical ideas and in the organization of math worlds.

THE SOCIAL LIFE OF PURE MATHEMATICS

Pure mathematics is grounded in and constituted out of social and material resources. The idea that pure mathematics is a product of some type of unmediated cognitive process is based on the difficulty of discovering the link between the thinking individual, social life, and the material world. It is just this discovery that is being slowly constructed on the foundations of the works of Durkheim, Spengler, Wittgenstein, and others. Establishing this link involves, in part, recognizing that symbols and notations are actually "material," and that they are worked with in the same ways and with the same kinds of rules that govern the way we work with pebbles, bricks, and other "hard" objects.

George Boole's attempt to discover the "laws of thought" failed because he did not understand the social and material bases of categorical propositions (Restivo, 1990:132ff.). Categorical propositions are actually high-level abstractions constructed out of "real world" experiences, grounded in the generational continuity of teacher-student and researcher-researcher chains that get reflected and expressed in chains of inductive inferences. The "self-evidence" of such propositions arises not from their hypothesized status as "laws of thought" but from their actual status as generalizations based on generations of human experience condensed into symbolic forms.

In the case of metamathematics, problems, symbols, and meanings that seem to be products of pure intellect and of arbitrary and playful creativity are in fact objects constructed in a highly rarified but nonetheless social world. Increasingly abstract ideas are generated as new generations take the products of older generations as the resources and tools for their own productive activities; still higher orders of abstraction are generated when mathematicians reflect on the foundations of abstract systems and self-consciously begin to create whole mathematical worlds.

We can watch this process of moving up levels of abstraction from the "primitive" ground or frame of everyday life in Boole's discussion of the "special law", $x^2 = x$. He begins by proposing in an abstract and formal way a realm of Number in which there are two symbols, 0 and 1. They are both subject to the special law. This leads him to describe an Algebra in which the symbols x, y, z, and c "admit indifferently of the values 0 and 1, and of these values alone." Now where does that special law come from? Boole actually constructs it on the basis of a "class" perspective grounded in "real world" examples such as "white things" (x), "sheep" (y), and "white sheep" (xy). This establishes that $xy = yx$. He also gets $1^2 = 1$ by interpreting "good, good men" to be the equivalent of "good men." In the case of $xy = yx$, he argues that since the combination of two literal symbols, xy, expresses that class of objects to which the names or qualities x and y represent are both applicable, it follows that "if the two symbols have exactly the same signification, their combination expresses no more than either of the symbols taken alone would do." Thus, $xy = x$; and since y has the same meaning as x, $xx = x$ (Boole, 1958:31, 37).

Finally, by adopting the notation of common algebra, Boole arrives at $x^2 = x$. We are now back in a world in which it is possible to make ordinary language statements such as "good, good men" and mathematical statements such as $1^2 = 1$. A careful review of Boole's procedure shows that far from creating a "weird" notation out of thin air, he simply describes a pristine "everyday" world in which only 0s and 1s exist. This is a rarified world indeed, but it is one in which the rules that govern the behavior of zeros and ones are similar to the ones that govern the behavior of material objects in the everyday material world. Symbols and notations are simply higher order materials that we work with the same way and under the same sorts of constraints that apply to "hard" materials. (Eventually, Boole interprets zero and one in logic as, respectively, Nothing and Universe).

It is interesting that there is a tendency for philosophers of mathematics and metamathematicians to reproduce a largely discredited naive realism to "explain" the process of operating on old abstractions and creating new ones. Kleene, for example, writes:

> Metamathematics must study the formal system as a system of symbols, etc. which are considered wholly objectively. This means simply that those symbols, etc. are themselves the ultimate objects, and are not being used to refer to something other than themselves. The metamathematician looks at them, not through and beyond them; thus they are objects without interpretation or meaning. (1971:63)

This process stylizes the idea of objective science. But without a sociological theory of intellect and knowledge, it is impossible to see that the same reasons for abandoning naive realism in physical science are relevant in the case of the so-called exact sciences (see the discussion in Restivo, 1990).

Some mathematicians and philosophers of mathematics have recognized that abstraction has something to do with iteration. They have expressed this recognition in such ideas as "second-generation abstract models," and "algebras constructed upon algebras." An important instance of invoking the iteration principle is Richard Dedekind's (1956:529) demand that "arithmetic shall be developed out of itself." One could even claim that mathematics in general is an iterative activity. Sociologically, iteration is the social activity of unbroken chains of mathematical workers, that is, generational continuity.

As generational continuity is extended and closure proceeds in a mathematical community, mathematicians work more and more in and less and less out of their math worlds. As a result, their experiences become progressively more difficult to ground and discuss in terms of generally familiar everyday world experiences. The worlds they leave behind are pictured worlds, landscapes of identifiable things. Math worlds are worlds of symbols and notations. This is the social and material foundation of so-called pictureless mathematics. But mathematical experiences in highly specialized math worlds are not *literally* pictureless. The resources being manipulated and imagined in math worlds are so highly refined that they are not picturable *in terms of everyday reality;* the referents for mathematical objects are increasingly mathematical objects and not objects from the everyday world. Since closure is never perfect, some degree of everyday picturing does occur even in the most abstract mathematical work; and in any case everyday pictures are almost inevitably produced as mathematicians move back and forth between math and other social worlds. At the same time, *new* picturing experiences and processes lead to the development of new pictures, math-world pictures. During the period that these pictures are being socially constructed (that is, while mathematicians are *learning* to interpret or "see" or "picture" objects in their math worlds), mathematical experience in its more abstract moments will necessarily appear pictureless.

THE MANUFACTURE OF PROOFS

In 1928, G. H. Hardy wrote:

There is strictly speaking no such thing as mathematical proof; . . . we can, in the last analysis, do nothing but point: . . . proofs are what Littlewood and I call gas, rhetorical flourishes designed to affect psychology, pictures on the board in lectures, devices to stimulate the imagination of pupils. (Kline, 1980:314)

Some years later, the mathematician R. L. Wilder argued that a proof is merely a way of "testing the products of our intuition":

Obviously, we don't possess and probably will never possess, any standard of proof that is independent of time, the thing to be proved, or the person or school of thought using it. And under these conditions, the sensible thing to do seems to be to admit that there is no such thing, generally, as absolute truth [proof] in mathematics, whatever the public may think. (Kline, 1980:314)

And Whitehead himself argued against grounding philosophic thought in the exact statements of special sciences; from the viewpoints of the advance of thought, exactness and logic are "fakes" (Kline, 1980:315).

The regard that mathematicians have for proof varies across time and space. J. Hoene-Wronski, Lacroix, and Jacobi were the mathematicians who contributed to the low regard for rigor and proof common in the 1850s (Kline, 1980:166). It is easy to find examples of a continuing discrimination against rigor and proof even in the "foundations"-sensitive twentieth century; witness the remarks by Hardy, Whitehead, and Wilder. Even Bertrand Russell wrote in 1903 that one of the chief merits of proofs is "that they instill a certain skepticism about the result proved" (Kline, 1980:315). Kline (1980:319) argues that in the end, logic was defeated in the rigorization and formalization of mathematics. As a result, proof once again takes on a supporting role, with intuition advancing as the fundamental basis for creative mathematics. Kline sounds a warning against the "logical fix" propounded by the Bourbakists.

The historical variations in the significance attributed to proofs and in the amount of energy devoted to proof work suggests that we should examine the relationship among variations (swings) and changes in the social structure of mathematical work. For the moment, let me simply note that proof work seems to take on increased, importance whenever mathematical work develops a certain level of social autonomy.

I want to consider whether we can conceive of proofs as cultural items and, in particular, whether we can view them as tools or machines for processing given mathematical objects and also for producing new mathematical objects. Kitcher gives a functionalist characterization of proofs as "sequences of sentences which do a particular job." By definition, he writes, a proof is "a sequence of statements such that every member of the sequence is either a basic a priori statement or a statement which follows from previous members of the sequence in accordance with some a priority-preserving rule of inference" (Kitcher, 1983: 37-38).

Proof anxiety is common in mathematics and is not a product of computer-assisted proofs. There is always the concern in reading a long proof that even if one is certain that each step is correct, an error may yet be hiding somewhere. Kitcher notes that there is a social psychology of proofs. Whether you are a beginner or an old hand, if others endorse your reasoning you become more convinced that your proof is correct. However,

> The reasonable doubt which arises when we follow complicated proofs can be exploited by circumstances in which we receive criticism rather than applause from the learned world. Reasonable uncertainty is typically compatible with knowledge because of the kindly nature of background experience. Transform the quality of our lives, and that knowledge could no longer coexist with the uncertainty. (Kitcher, 1984:41-43)

Some intuitionists have suggested that all a proof does is *verify* that certain constructions have certain properties (Kitcher, 1983:143). But other accounts suggest other functions for proofs (Kitcher, 1984:180-81): for example, they can yield new knowledge, or help us obtain new knowledge from old knowledge. But if and where a proof process is in fact "a well-defined repetitive transformation of formulas into new formulas," we can mechanize the process (Hawkins, 1964:34). This is a highly suggestive remark and prompts me to consider to what extent proofs, in general, are machines.

Curry (1951) conceives formal systems as defined by a primitive frame, that is, a set of conventions. This primitive frame gives "the engineer" (metamathematician or mathematician) "all the data he needs (apart from his knowledge of engineering) for constructing the required formula-producing machine." This seems to be consistent with definitions of formal proofs such as the following:

> "A (formal) proof is a finite sequence of one or more (occurrences of) formulas such that each formula of the sequence is either an axiom or an immediate consequence of preceding formulas of the sequence" (Kleene, 1971:83).

Consider a proof as an *experiment*. We set up an apparatus called the "Given" or the "If . . ." We then "run the experiment." In the proof, we carry out, intuitively and/or constructively, operations allowed by the apparatus—or, we allow the apparatus to "run." We then look for an "effect." This is called the "Then . . ." or the "It follows that . . ." This notion of proof, incidentally, follows immediately from the conception of proof as demonstration.

ARE MATHEMATICAL OBJECTS REAL?

The reality of mathematical objects is exactly what one sees on successive pieces of paper (or in any medium in which we can make mathematical objects manifest). In this sense, Hilbert's formalism was correct. But the marks on the paper are not arbitrary. They are historically linked to earlier notations and meaning. Mathematics progresses by taking operations formerly treated concretely (such as a specific function, a particular number) and creating a representation or symbol for them. In this way, what was formerly either concrete or intuitive becomes available as a distinct object *on paper*. This process is the basis for both crude empiricist *and* Platonist perspectives in mathematics. Chains of abstract or higher mathematics *can* be traced back to representations of the physical world, hence the empiricist position. The representations, meanwhile, have their own special properties, properties that are not found in the physical world. These properties are symbolized and utilized in mathematical operations. The representations take on a "thinglike" quality, hence Platonism. Let me try to make these ideas a little more concrete.

Consider Kitcher's recommendation that we think of mathematical objects in terms of mathematical activities. Thus, we should think not of a *collection* but of *collecting* (Kitcher, 1983:110). Elementary collecting involves the physical manipulation of material objects. Experience with collecting eventually makes it possible for us to engage in thoughtcollecting. We collect objects by "running through a list of names, or by producing predicates which apply to them;" a hierarchy of collectings can be produced by assigning symbols to the products of a given collecting, and repeating collecting operations on those symbols: "notation makes it possible for us not only to talk (e.g.) about functions from objects to objects (which correspond to certain first-level correlations) but also about functions from functions to functions, and so forth" (Kitcher, 1983:111). Collating or matching material objects is another elementary mathematical activity. Note that Kitcher does not contend that higher-order collecting involves collecting *symbols*. Rather, *prior operations* are collected. Symbols, of course, are used in carrying out operations: "To collect is to achieve a certain type of representation, and, when we perform higher order collectings, representations achieved in previous collecting may be used as materials out of which a new representation is generated." (Kitcher, 1983:129; cf. Restivo and Collins, 1982).

Kitcher's model for the notion of operating on operations comes from set theory. Following Boolos (1971:215-31), Kitcher (1983:131) ar-

gues that sets are formed in stages. The procedure is: (1) form all sets
of individual elements: (2) form all collections of individuals and sets
in (1); (3) at the $(n + 1)$th stage, form all collections of individuals and
sets formed up to and including the nth stage. This procedure can be
continued indefinitely, into the transfinite realm. The Zermelo-
Fraenkel set theory hierarchy is constructed through the iterated formation of sets out of previously produced mathematical "material."
Kitcher contends that what is true of set theory can be used to model
any mathematical activity. His case study of the calculus is designed
to illustrate the viability of the set theory model of operating on operations. In fact, he is concerned with showing that changes in the history of the calculus were "rational" (see my interpretation of this
move in the next section). He does not give an explicit translation of
calculus activity into interated sets terms. However, he does offer numerous examples of iteration. The idea of iteration as a basic mathematical process has been noted by many mathematicians, as I pointed
out earlier.

The two most important ideas Kitcher uses in discussing the concept of operating on operations are 'iteration' and 'referent'. The latter
concept is used in the sense of identifying the referents for mathematical objects. Kitcher shows that (as we might expect from my earlier
discussion of mathematical work) the referents are either mathematical
objects or mathematical objects representing material objects or other
mathematical objects. For example, consider the referent for "number."
In the late nineteenth century, the referent for number might have
been fixed by saying that anything denoted by an expression "$a+ib$"
(where a and b are decimal expressions) is a number. Given this way
of fixing reference, Cantor's concept of a number that follows all the
natural numbers would "not refer." But this would not be the case if
we fixed the referent for number in terms of operations. So, if we say
that numbers are just those things that we can add, subtract, multiply,
and divide, then Cantor's transfinite objects can be shown to be numbers if it can be shown that they are objects that can be added, sub-
tracted, multiplied, and divided. Note that Cantor explicitly linked
his theory of transfinite number to the theory of infinite point sets and
a transfinite *arithmetic*: "there are paradigms of succession in which all
the numbers of a series are succeeded even though the series has no
last number" (Kitcher, 1983:174). For example, 1 follows all the members of the infinite series $< 1/2, 3/4, \ldots, 1 - 1/2n, \ldots >$.

If we fix the referent of "number" by using paradigms such as 3,
1, 2, and so on, that is, if we restrict reference to *real* numbers, then -1
fails to refer. But if we conceive of number in terms of operation, then

−1 does refer. Bombelli and others fixed the referent for number, in effect, in the latter way and thus could allow expressions such as −1 to enter their calculations: "What needed to be done to show that the more restrictive mode of reference should be dropped from the reference potential of 'number' was to allay fears that recognizable analogs of ordinary arithmetical operations could not be found" (Kitcher, 1983:175). This led to fixing reference so that the reference potential of −1 "came to include only events in which *i* was identified as the referent." This required showing that there were parallels between *complex* and *real* numbers in terms of susceptibility to arithmetical operations.

Natural numbers (natural number operations) are grounded in counting. Real numbers (real number *operations*) are grounded in measurement. Bombelli was in effect pointing to "hitherto unrecognized operations." Having shown that the new operation would submit to "recognizably arithmetical treatment," the next step was to show that the new operation could be "construed in physical terms." This was accomplished by showing that $a + bi$ referred to the operation of vector displacement. Complex multiplication referred to the operation of rotation. This is a framing process (Goffman, 1974). Thus, in the conventional frame in use before Bombelli's work, there was consensus that "there is no number whose square is −1." Bombelli broke frame, and reframed, so that now what was "true" was that "there is no *real* number whose square is -1" (Kitcher, 1983:179).

Let us return now to the process of iteration. Following Kitcher, [{*a,b*}, {*c,d*}] can be viewed as a sequence of collecting operations. First, we collect *a* and *b*; then we collect *c* and *d*; and then we operate on these collectings. "{ . . . }" represents first-level collecting; iterating this notation is a new level of collecting activity. Formally, Kitcher introduces the following *primitive notions*: (1) one-operation; (2) successor (one operation being a successor to another); (3) addition (one operation being an addition on other operations); (4) matchability of operations:

> We perform a one-operation when we perform a segregative operation in which a single object is segregated. An operation is a successor of another operation if we perform the former by segregating all of the objects segregated in performing the latter, together with a single extra object. When we combine the objects collected in two segregative operations on distinct objects we perform an addition on those operations. (Kitcher, 1983:111-12)

Axioms and definitions "systematize previously accepted problem solutions. Foundational study is motivated by the need to fashion tools for continuing mathematical research" (Kitcher, 1983:271). What *I* have tried to make *explicit* is the organizational grounds and implications of precisely a philosophical view such as Kitcher defends, a philosophical view that has been infected but not overcome by social and cultural ideas.

Let me now offer a more mundane portrait of iteration that helps to make clear its nature as a form of work. This portrait will also recapitulate the discussion of interation in a way that relates it to my earlier conjectures about specialization and closure.

Let us begin by entering the Level One world. This world contains discrete objects, such as APPLES and TREES (keep in mind that we are already at a certain level of abstraction but one which can serve as a starting point for us). Some objects in the Level One world are linked together by other objects. For example, STEMS link APPLES and TREES. Cultural development leads to the construction of models of the objects we encounter and use in the Level One world, and one of the consequences of carrying out this process is that we abstract qualities, smooth over rough edges, and idealize the Level One objects (I bracket the complex reasons for these moves, but clearly they are related to human activities in the service of survival, growth, understanding, and control). This process can be illustrated using a Tinker Toy model.

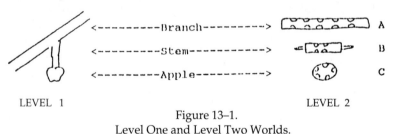

LEVEL 1   LEVEL 2

Figure 13–1.
Level One and Level Two Worlds.

In figure 13-1, we construct models of the Level One world and create a Level Two world. Simultaneously, we add three new objects to the Level One world. A fourth object is added when we complete the construction in figure 13-2. We can now construct additional objects such as AB, AC, AD, and so on. And we can reproduce A, B, C, and D once or many times (depending, of course, on the availability of the relevant resources), and then create more complicated constructions (E, for example, by combining two Ds). These latter moves, notice, are not directly related to the Level One world. In specialization,

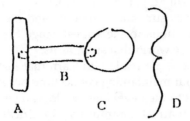

Figure 13–2.
Construction of an Abstract Object.

we become increasingly involved in Level Two and higher-level worlds.

With the foregoing in mind, consider then that in mathematics, the world of three-dimensional Euclidean geometry is constructed in and out of the space immediately around and accessible to us with its points, lines, planes, spheres, and so on (again, keep in mind that we are already operating at a level of abstraction at this stage). The set theory world would consist of the so-called naive sets and naive membership relations. The number theory world includes the natural numbers, the successor function, the operations of addition and substraction, and so on (Snapper, 1979:552).

What happens if we carry on this iterative process indefinitely? Up to a certain point, we will be able to come back to the world of trees and apples and say something useful about that world. The reason is that the iterations actually incorporate aspects of the Level One world we are not aware of, and the model play allows us to find out about some of these aspects. But at some point, we will move beyond the capacity of higher-level worlds to communicate with the Level One world. We may find, at some point, that we can't get back to Level Two; then Level Three will be our limit, and so on. As iteration continues, in conjunction with the constraints of increasingly narrow forms of social organization, the range of worlds we can traverse rapidly decreases.

Specializing in and "purifying" mathematics can become a process of adapting to narrower and narrower conditions. It can become, like other generalizing and specializing processes that occur in cultural systems, a dead-end strategy. The result of such a strategy would be somewhat analogous to the withdrawal of the person into him- or herself until the self-system collapses on itself (a form of this sort of process is what we see at work in certain types of mental illness). This analogy suggests the importance of interplay between the centripetal forces of closure and autonomy and the centrifugal forces that maintain connections to the widest range of worlds and especially to the Level One world that sustains us.

It may surprise some readers to learn that just the ideas I have been discussing about maintaining connections across worlds, and especially connections to the Level One world, can be found in the literature of the mathematical community. Specifically, some applied mathematicians have argued that pure mathematics has gone so far that the possibilities realized in the past for contributing to our work in the Level One world are rapidly diminishing (Boos and Niss, 1979).

Each successive form of mathematics is grounded in communities of mathematicians competitively solving problems and collectively judging what are to be accepted as valid solutions. The most "practical" and "empirical" (in the conventional sense of these terms) forms of mathematics are also the forms that inhibit the autonomy of the mathematical community. The degree to which mathematicians can abstract within these forms, the degree to which they can make metaobjects out of their own operations, is limited by the concrete frame of reference within which these forms are developed and used. Higher level abstractions—pure mathematics—depend upon mathematicians becoming increasingly self-enclosed organizationally (as a "community"), becoming more specialized and autonomous (in the sense noted above), and more professionalized in successive stages. It is also necessary to develop and maintain strong linkage across generations. This gives rise to a growth in the nested self-reference of mathematics.

Each successive level of abstraction (symbolically and notationally) embodies the kinds of operations previous generations of mathematicians used. But by "embodying" the earlier operations, it in a sense "kills" or "embalms" them (cf. Spengler, 1926). The operations so treated are no longer open; they become fixed, thinglike, and unproblematic (this is probably more familiar for the cases of machines and instruments). What appears to be a sort of "transcendental objectivity" in mathematics is really a matter of the extremely secure social foundations of the mathematical community.

The social aspects of mathematics, then, have two dimensions: *historical* (in the embodiment of past operations now reduced to thinglike representations, tools, and machines) and *contemporary* (each mathematician is implicitly and explicitly negotiating with—cooperatively or conflictually—his/her community over what problems are worthy of attention, what methods are appropriate, and what solutions will be acceptable). The latter constitutes the content of mathematical "intuition." As Goodman (1979) points out, a mathematician's intuition must be different from the formalism, because the intuition is a preformal idea for which a formalism is eventually sought (as the end result of the process of articulating and mechanizing intuition). Intuitions are not irreducibly mysterious—they are mathematical ideas still under negotiation. This "cutting edge" will be there as long as there is the generational continuity that allows mathematicians to build on previous issues and notations, and as long as there is insulation and competition.

The embodied part of mathematics is precisely the part that one no longer has to negotiate or struggle with; one simply takes previous operations for granted. In a sense, then, even though mathematics may be unique in the extent to which it realizes Whitehead's dictum—that science that hesitates to forget its founders is lost—the ancients are still *used* in the embodied forms. And of course one can still negotiate and struggle with dead mathematicians in one's own mind.

THE END OF EPISTEMOLOGY AND
PHILOSOPHY OF MATHEMATICS

The sociological imperative is having an impact on the work of mathematicians and philosophers who utilize the vocabularies of psychology, culture, empiricism, and pragmatism. But there is some resistance to giving full expression to that imperative. Take, for example, Philip Kitcher's views on the nature of mathematical knowledge. Kitcher, as an empiricist epistemologist of mathematics, constructs a "rational" explanation of beliefs and knowledge that brings psychology into the philosophy of mathematics, but in its psychologistic form. But psychologism cannot carry the burden of his attack on the apriorists unless it is recognized for what it is—a truncated sociology and anthropology. Kitcher seems to realize this at some level. He understands that knwledge has to be explained in terms of communities of knowers and that stories about knowledge can be told in ways that reveal how knowledge is acquired, transmitted, and extended. This is the *only* story Kitcher can tell; but he is intent on making his story confirm rationality and well-founded reasoning in mathematics.

Rationality and well-founded reasoning cannot be separated from social practice and culture. Where it appears that we have effected such a separation it will turn out that we have simply isolated mathematical work as a sociocultural system and told a sociologically impoverished story of how that system works. The extent to which mathematics is an autonomous social system will vary from time to time and place to place, and so then will the extent to which an empiricist epistemologist can construct a rational explanation for mathematics. "Rational" refers to the rules governing a relatively well organized social activity. Taking the sociological turn means recognizing that *rational* is synomous with *social* and *cultural* as an explanatory account. Kitcher can save the rationality of mathematics only by showing (as he does very nicely) that mathematics is a more or less institutionally autonomous activity. But this makes his so-called rational account nothing more than a social and cultural account. He misses this point in great part because he thinks primarily in psychologistic terms. As soon as we replace psychologism with the sociological imperative, rationality (as a privileged explanatory strategy) and epistemology (as a philosophical psychologistic theory of knowledge) are nullified.

The same situation is characteristic of philosophical treatments of knowledge and perception in general which are semiconscious of the sociological imperative but repress its full power. Richard Rorty's (1979) pragmatism starts out as a strategy for extinguishing epistemology; but in the end, Rorty Westernizes (or, more precisely, Americanizes) epistemology (this a reflection of the power of pragmatism as an "American philosophy") and gives it a reprieve. He explicitly restricts moral concern to "the conversation of the West." On the brink of a radical social construction conjecture on the nature of knowledge, he is restrained by (1) his stress on the ideal of polite conversation and his failure to deal with the more militant and violent forms of social practice in science and in culture in general; (2) his Western bias, manifested in his intellectual and cultural debts; (3) his Kuhnian conception of the relationship between hermeneutics ("revolutionary inquiry") and epistemology ("normal inquiry") which is prescriptive, an obstruction to critical studies of inquiry, and the coup that saves epistemology; and (4) his focus on an asociological conception of justification. Patrick Heelan (1983) also shackles a potentially liberating contextualist theory (this time of perceptual knowledge) by talking about Worlds belonging to the Western community even while he implicates himself in the project of the "redemption" of science from its Babylonian captivity in the West.

And finally, I recall Bloor in this context and his strong program, which helped set the stage for the study of mathematical and scientific knowledge by the new sociologists of science. He dilutes the sociological imperative with a normative commitment to Western culture and Western science.

Given so much Westernism in philosophy, I have to wonder about the extent to which philosophers such as the ones I have been discussing are ideologues of cultural orthodoxy and prevailing patterns of authority. The reality of these sorts of ideological chains is exhibited in the ways in which philosophers, pressured by empirical and ethnographic research to see that Platonism and apriorism (along with God) are dead, reach out adaptively for social construction. Their failures are tributes to the professionalization process in philosophy and its grounding in psychologism. This is all relevant for an appreciation of the significance of the sociological imperative in math studies because mathematics is "the queen of the sciences," the jewel in the crown of Western science. The protective, awe-inspired, worshipful study of mathematics is thus understandable—readily as a defense or appreciation of Western culture, less readily as a vestigial homage to the Western God. In either case we are closer to theology than to sociology of mathematics.

CONCLUSION: VALUES AND THE SOCIOLOGICAL IMPERATIVE

Math worlds are social worlds. But what kinds of social worlds are they? How do they fit into the larger cultural scheme of things? Whose interests do math worlds serve? What kinds of human beings inhabit math worlds? What sorts of values do math worlds create and sustain? In his description and defense of "the sociological imagination" (a form of the sociological imperative) C. Wright Mills (1961) drew attention to the relationship between personal troubles and public issues, the intersection between biography and history in society, and questions about social structure, the place of societies in history, and the varieties of men and women who have prevailed and are coming to prevail in society. If we approach math worlds from this standpoint, the questions we ask will be very different from those that philosophers, historians, and sociologists usually ask. The questions I have posed elsewhere concerning science worlds in general apply to math worlds too:

> . . . What do scientists produce, and how do they produce it; what resources do they use and use up; what material by-products and wastes do they produce; what good is what they produce, in what social contexts is it valued, and who values it; what costs, risks, and benefits does scientific work lead to for individuals, communities, classes, societies, and the ecological

foundations of social life . . . What is the relationship between
scientists and various publics, clients, audiences, patrons; how
do scientists relate to each other, their families and friends, their
colleagues in other walks of life; what is their relationship as
workers to the owners of the means of scientific production;
what are their self-images, and how do they fit into the communities they live in; what are their goals, visions, and motives?
(Restivo, 1988:218)

These questions are relevant to the study of math worlds because they
help us to recover the social worlds that get progressively excised in
the process of producing and finally presenting (and re-presenting)
mathematical objects.

Explaining the "content" of mathematics is not a matter of constructing a simple causal link between a mathematical object, such as
a theorem, and a social structure. It is rather a matter of unpacking
the social histories and social worlds embodied in objects such as theorems. Mathematical objects are and must be treated *literally* as objects, *things* that are produced by, manufactured by social beings.
There is no reason that an object such as a theorem should be treated
any differently than a sculpture, a teapot, or a skyscraper. Only alienated and alienating social worlds could give rise to the idea that
mathematical objects are independent, free-standing creations and
that the essence of mathematics is realized in technical talk. Notations
and symbols are tools, materials, and, in general, resources that are
socially constructed around social interests and oriented to social
goals. They take their meaning from the history of their construction
and usage, the ways they are used in the present, the consequences of
their usage inside and outside of mathematics, and the network of
ideas that they are part of. The sociological imperative, especially
when informed by the sociological imagination, is a tool for dealien-
ation and for uncovering the images and values of workers and social
worlds in mathematics.

APPENDIX 1: EMPIRICAL PHILOSOPHY OF MATHEMATICS

It is useful to acknowledge philosophers of mathematics who
treat mathematics as an empirical science, or what I would call more
generally a "culture." L.O. Katsoff, for example, favorably cites Gron-
seth's remark that logic is the physics of objects per se. This is a remark that could be applied to formal mathematics or even pure math-

ematics, perhaps—if these distinctions even make sense. According to Katsoff, who was influenced by Husserl: "Mathematics is not merely a body of symbols, but is a construct of humans which is evolved in a society and which has relation to empirical events. It therefore has syntactical, sociological, psychological, and scientific fields of investigation" (Katsoff, 1948:253).

An earlier, more primitive example of mathematical work as analogous to work in general is provided by de Morgan. He dealt with the "mathematization of mathematics" as a correlate of professionalization. Augustus de Morgan was a "satellite" of the Analytical Society. He was dedicated to the notion of mathematical work as a profession. At the same time, he stressed the importance of a specifically mathematical symbolism as something that could suggest new results.

In teaching, we might use examples such as the following:

$$3^2 - 2^2 = (3 - 2)\ (3 + 2)$$
$$6^2 - 4^2 = (\quad)\ (\quad)$$
$$a^2 - b^2 = (\quad)\ (\quad)\ a > b$$
$$a^2 - b^2 = (\quad)\ (\quad)$$

In this set of steps, we recognize an orientation to *generalizing* a number fact to numbers in general. Consider the following more strictly mathematical example:

$$(a + b)^2 = a^2 + 2ab + b^2$$
$$(a - b)^2 = \ldots$$

.

.

.

$$(a \pm b)^n = \ldots \text{ for any integer n}$$

Again, these are steps in generalization. And a concern for these sorts of steps, especially at the more advanced stages of development, is a sign of professionalization. Thus, de Morgan's self-consciousness about the need to extend meaning, about the interpretive aspect of math work, occurs as a part of or correlate of the general development of mathematics as a profession. An especially interesting example of de Morgan's approach is the following:

1. $2X = X$ (for $X = 0$)
2. $2X/X = X/X$
3. $2 = 1$

This bit of mathematical magic, intended as a manipulation of old mathematics in the interest of creating a new mathematics, was sterile. But when Boole carried out a similar move, from $(1)(1) = 1$ and $(0)(0) = 0$ to $(X)(X) = X$, he set the foundation for a highly productive algebra.

According to Mannoury, mathematics is a "phenomenon of life" (*Lebenserscheinung*). He views mathematics from the perspective of "significs," "the theory of the mental associations which lie behind human verbal acts but exclusive of the sciences of language in the narrower sense (semantology, etymology, linguistics and philology)" (Katsoff, 1948:197). Mathematics is made up of two parts: the words and symbols it manifests itself in (formalistic mathematics); and the "system of mental association" on which the formalistic mathematics is based. According to Mannoury:

> A mathematical formula, apart from mere study purposes, is only expressed or written down when it is to be applied to some experiential datum. The purpose of this application is not determined by the formula but by the relation of the person speaking to the object of his calculations. The formula, or more generally, the more or less "pure" mathematics, is only the form of appearance of the transfer from purpose to means. (Katsoff, 1948:199)

Mannoury then examines the nature of natural laws and mathematical statements from a psychologistic perspective. Here, as in other cases where the human grounding of science or mathematics is recognized, "psychology" is readily recognized to imply a broader social and cultural perspective. Natural laws, he argues, cannot have any objective content in any strict sense, that is, a content independent of human life: they are, in the end, psychological laws (bracketing the problem of "laws," I would argue for the term sociological here). These laws are:

> only expressions for regularities which the associations between remembered expectations and experiences disclose (the content of experience or indicative meaning of natural law) on the one hand, and on the other for the corresponding regularities in the associations between experience and expectation (emotional-volitional meaning of natural law) (Katsoff, 1948:200).

Now Mannoury claims that there is a distinction between formal mathematical and formal physical linguistics or verbal acts:

The mental work of the most isolated mathematician consists of linguistic acts. Therefore it can and must be tested in the indicative and emotional meaning elements. It is clearly evident that even this type of mental work is conditioned by memory and expectation associations, as is every other type. Only here the "experiential content" of the mathematical theorem or proof consists in the knowledge common to the author and reader (Katsoff, 1948:202).

But in principle we are dealing here with two arenas for generating "experiential content," and not two fundamentally different verbal acts. Indeed, even where Mannoury tries to defend the distinction between natural and mathematical laws he concludes that "physical and verbal regularities mutually influence each other": "Thus the 1 + 1 = 2 and the 'natural law' of persistence are closely tied together" (Katsoff, 1948:204). There are other inconsistencies in Mannoury, or in Katsoff's interpretation of Mannoury. For example, according to Katsoff, Mannoury "affirms that we can speak only of formalistic formal-mathematical language when the linguistic reaction of the hearer to the verbal act of the speaker is, as much as possible, independent of the persons and of the accompanying phenomena" (Katsoff, 1948:205).

But this doesn't seem to be congruent with Katsoff's (1948:206) summary remarks in which Mannoury is said to claim that "mathematics is relative to linguistic acts and verbal forms," that its development is "relative to the development of human language and thought," and that it "cannot be independent of human beings and their purposes." Perhaps the inconsistencies reflect Mannoury's psychologistic perspective, an advance on purist views but an abortive sociological approach. It should be noted that for Mannoury, "the decisive method to use in considering the foundations of mathematics is the *empirical* and especially the *psychological* method. Thus significant or psycholinguistic investigation must precede axiomatic investigations" (Katsoff, 1948:206). This "order of analyses" notion is another possible clue to the inconsistencies in Mannoury's approach.

Given the preceding background of this chapter, we can see why it is not a simple matter to argue (as Körner [1926:174] does) that the grounds for "There exists a piece of copper" and "There exists a Euclidean point" are "quite different." Indeed, this is an example of what I have called elsewhere the "fallacy of the incommensurable comparative." On the one hand, the copper statement is rooted in a primitive relationship to some object in the physical world for Körner.

This ignores the historical development of the concept of copper and the rarification of the abstract idea, 'copper', within, for example, metallurgy. On the other hand, the concept of a Euclidean point is the current culmination of an abstractive process that begins with statements in the world I level of the copper statement.

APPENDIX 2: POLITICS, EDUCATION AND MATHEMATICS

Mathematics has been a tool of ruling elites and their political opponents from its development in the ancient civilizations until the present. In modern history, we can point to the examples of Napoleon I asserting that "The advancement and perfection of mathematics are intimately connected with the prosperity of the State" (Moritz, [1914] 1942:42); and the slogan adopted in Mozambique in the 1970s, "Let us make mathematics a weapon in the building of socialism" (Zaslavsky 1981:16). Mathematics, science, and knowledge in general are crucial resources in all societies. Systems of knowledge therefore tend to develop and change in ways that serve the interests of the most powerful groups in society. Once societies become stratified, the nature and transmission of knowledge begin to reflect social inequalities. Today, we see this principle illustrated in the mathematics curriculum: "it is the content and methodology of the mathematics curriculum that provides one of the most effective means for the rulers of our society to maintain class divisions" (Zaslavsky, 1981:15).

The mathematical curriculum, like curricula in general, is conditioned by the social functions of education in a stratified society. Educational institutions in advanced industrial societies "foster types of personal development compatible with the relationships of dominance and subordinancy in the economic sphere": "The rule orientation of the high school reflects the close supervision of low-level workers; the internalization of norms and freedom from continual supervision in elite colleges reflect the social relationships of upper-level white-collar work. Most state universities and community colleges, which fall in between, conform to the behavioral requisites of low-level technical, service, and supervisory personnel." (Bowles and Gintes, 1976:11-12.) This form of political economy fosters low achievement expectations among teachers in inner-city schools and allows children to be promoted on the basis of "good conduct." In the wealthy districts, meanwhile, the schools "can offer up-to-date curricula, well-trained teachers, small classes, and computer education for every student" (Zaslavsky, 1981:20). In this context, the "back to ba-

sics" movement in mathematics (and other curricula) can be understood as a means for (1) cutting funds for education and "freeing resources for an even greater military buildup"; and (2) persuading "the
poor, the minorities, and females to accept their inferior status in a
capitalist society" (Zaslavsky, 1981:24). The same sort of situation is
reproduced in the international system of stratification (D'Ambrosio,
1985).

The relationship between mathematics, class, and power in the
everday world is given deeper meaning by the sociology of mathematics I discussed earlier. The "social construction" perspective
shows *how* deeply politics, education, and other social factors are im-
plicated in mathematical work and mathematical knowledge. One important implication of the constructivist perspective is that mathematical reforms (or more radical change) cannot be effectively carried out
in isolation from broader issues of power, social structure, and values.
If, on the other hand, we adopt conventional mathematical tools and
ways of working to help solve social, personal, and environmental
problems, we will fall short of our goals. As a social institution, modern mathematics is itself a social problem in modern society. It is
therefore unreasonable to suppose that social reformers and revolutionaries could *eliminate* mathematics from society; and equally unreasonable to suppose that mathematical reformers or revolutionaries
could *force* mathematics as we know it today into some "alternative"
shape independently of broader social changes.

What, then, should our approach to social change in mathematics and society be? The constructivist perspective suggests that we
should focus on transforming ways of living, social relationships, and
values in society at large. As I have noted elsewhere, "A radical
change in the nature of our social relationships will be reflected in
radical changes in how we organize to do mathematics—and these
changes will in turn affect how we think about the content of our
mathematics" (Restivo, 1983:266).

There are a number of ways in which modern mathematics can
be considered a social problem. It tends to serve ruling-class interests;
it can be a resource that allows a professional and elite group of mathematicians to pursue material rewards independently of concerns for
social, personal, and environmental growth, development, and well-
being; aesthetic goals in mathematics can be a sign of alienation or of
false consciousness regarding the social role of mathematicians; and
mathematical training and "education" may stress "puzzle solving"
(in Kuhn's sense) rather than "ingenuity, creativity, and insight"
(D'Ambrosio 1985:79). Efforts to address these problems will fail if

they are based on the view that mathematics is a set of statements, a body of knowledge, or a methodology. This view reflects the notion that *technical* talk about mathematics gives us a complete understanding of mathematics. If, on the other hand, we adopt the constructivist perspective that *social* talk about mathematics is the key to understanding mathematics (including mathematical knowledge), then our approach to solving the social problems of mathematics *and* the problem of "mathematics as a social problem" will necessarily focus on social roles and institutions.

New social circumstances and arrangements will give rise to new conceptions and forms of mathematics. We cannot anticipate these new conceptions and forms; to a large extent, we cannot even imagine them. We *can* imagine practicing, teaching, producing, and using mathematics in new ways. This does not require attacking all social ills at all levels simultaneously; it does, at the very least, require that we approach revisions, reforms, and revolutions in mathematics always with an awareness of the *web* of roles, institutions, interests, and values mathematics is imbedded in and embodies.

NOTE

This chapter is based on my contributions to the special issues of *Philosophica* and *ZDM*. Some of the later sections are based on my work in the sociology of mathematics with Randall Collins, including some work presented at the 1984 meeting of the Society for Social Studies of Science in Gent, Belgium in my lecture, "The Social Realities of Mathematical Knowledge."

REFERENCES

Becker, H.
   1982 *Art Worlds*. Berkeley and Los Angeles: University of California Press.

Bloor, D.
   1976 *Knowledge and Social Imagery*. London: Routledge and Kegan Paul.

Boole, G.
   1958 *The Laws of Thought*. Reprint; originally published 1854. New York: Dover.

Boolos, G.
   1971 "The Iterative Conception of Set." *Journal of Philosophy* 68:215-231.

Booss, B. and M. Niss (eds.)
   1979 *Mathematics and the Real World*. Boston: Birkhouser.

Bottomore, T., and M. Rubel, eds.
1956 *Karl Marx: Selected Writings in Sociology and Social Philosophy.* Translated by T. Bottomore. New York: McGraw-Hill.

Bowles, S. and H. Gintis
1976 *Schooling in Capitalist America.* New York: Basic Books.

Collins, H.
1985 *Changing Order.* Beverly Hills: Sage.

Crumb, T.
1989 *The Anthropology of Number.* Cambridge: Cambridge University Press.

Curry, H. B.
1951 *Outlines of a Formalist Philosophy of Mathematics.* Amsterdam: North-Holland.

D'Ambrosio, U.
1985 *Socio-cultural bases of mathematical education.* Campinas, Brazil: UNI-CAMP.

Dedekind, R.
1956 "Irrational Number." Pp. 528-36 in J. R. Newman (ed.), *The World of Mathematics,* vol. 1. New York: Simon and Schuster.

Durkheim, E.
1961 *The Elementary Forms of the Religious Life.* New York: Collier Books.

Fleck, L.
1979 *Genesis and Development of a Scientific Fact.* Translated by F. Bradley and T. Trenn. Chicago: University of Chicago Press.

Geertz, C.
1983 *Local Knowledge.* New York: Basic Books.

Goffman, E.
1974 *Frame Analysis.* New York: Harper Torchbooks.

Goodman, N. D.
1979 "Mathematics as an Objective Science." *American Mathematical Monthly* 86, 7:540-557.

Gumplowicz, L.
1905 *Grundrisse der Soziologie.* Vienna: Manz.

Hawkins, D.
1964 *The Language of Nature.* San Francisco: W. H. Freeman.

Heelan, P.
1983 *Space-Perception and the Philosophy of Science.* Berkeley and Los Angeles: University of California Press.

278   *Mathematics, Society, and Social Change*

Katsoff, L. D.
1948 *A Philosophy of Mathematics.* Ames: Iowa State College Press.

Kitcher, P.
1983 *The Nature of Mathematical Knowledge.* New York: Oxford University Press.

Kleene, S.
1971 *Introduction to Metamathematics.* New York: American Elsevier.

Kline, M.
1980 *Mathematics: The Loss of Certainty.* New York: Oxford University Press.

Körner, S.
1962 *The Philosophy of Mathematics.* New York: Harper Torchbooks.

Latour, B.
1987 *Science in Action.* Cambridge, Mass.: Harvard University Press.

Mills, C. Wright
1961 *The Sociological Imagination.* New York: Grove

Moritz, R. E.
1942 *On Mathematics and Mathematicians.* New York: Dover.

Restivo, S.
1983 *The Social Relations of Physics, Mysticism, and Mathematics.* Dordrecht: D. Reidel.
1988 "Modern Science as a Social Problem." *Social Problems* 35:206-25.
1990 "The Social Roots of Pure Mathematics." Pp. 120-43 in S. Cozzens and T. Gieryn (eds.), *Theories of Science in Society.* Bloomington: Indiana University Press.

Restivo, S. and R. Collins
1982 "Mathematics and Civilization." *The Centennial Review* 26, 3:277-301.

Rorty, R.
1979 *Philosophy and the Mirror of Nature.* Princeton: Princeton University Press.

Snapper, E.
1979 "What is Mathematics?" *American Mathematical Monthly* 86:551-557.

Spengler, O.
1926 *The Decline of the West.* Vol. 1. New York: International Publishers.

Zaslavsky, C.
1981 "Mathematics Education: The Fraud of 'Back to Basics' and the Socialist Counterexample." *Science and Nature* No. 4:15-27.

Contributors

Jean Paul Van Bendegem teaches at the Vrije Universiteit in Brussels, Belgium, and is the editor of the journal *Philosophica*. His publications span the fields of logic, mathematics, science policy, and philosophy. He is the author of many papers, including "Fermat's Last Theorem seen as an Exercise in Evolutionary Epistemology" (1987), and "Non-formal Properties of Real Mathematical Proofs" (1988). His book, *Finite, Empirical Mathematics: Outline of a Model*, was published in 1987.

Philip J. Davis is professor of Applied Mathematics at Brown University and author of *The Schwarz Function* (1974), *Circulant Matrices* (1979), and other works in numerical analysis and approximation theory. His book *The Mathematical Experience*, coauthored with Reuben Hersh, was awarded the American Book Award for 1983. His many honors in mathematics include the George Polya Award (1987) and the Hedrick Award (1990).

Roland Fischer is professor of Mathematics at the Univesity of Klagenfurt in Klagenfurt, Austria. He has a special interest in didactics and has been involved in the development of an in-service-training course for mathematics teachers. He is the scientific adviser to the Austrian Ministry of Education on curriculum planning in mathematics. Dr. Fischer has published widely on mathematics and mathematics education and he is the coauthor of *Mensch und Mathematik* (1985).

Helga Jungwirth is a free-lance scientist who specializes in mathematics education. Her main research topics are gender and mathematics and the use of computers in the classroom. She studied mathematics and physics at the Johannes Kepler University in Linz, Austria, worked as a secondary school teacher for a number of years, and is now involved in in-service teacher training.

Herbert Mehrtens teaches the history of science at the Technische Universität Berlin and has done research on the history of technology and of mathematics and natural science in Germany. He is the author of *Die Entstehung der Verbandstheorie* (1979) and *Moderne-Sprache-Mathematik* (1990), a coeditor of *Naturwissenschaft, Technik und NS-Ideologies* (1980), and a coeditor of *Social History of Nineteenth Century Mathematics* (1980).

Nel Noddings teaches in the Department of Education at Stanford University. She is active in research on mathematics education and is the author of *Caring: A Feminine Approach to Ethics and Morals* (1984).

Yehuda Rav is a mathematician, philosopher, and educator and teaches at the Université de Paris-Sud. Before joining the mathematics faculty there in 1967, he taught at Columbia University and Hofstra University in the United States. He is the author of more than fifty publications in mathematics, logic, cybernetics, and philosophy.

Sal Restivo is professor of Sociology and Science Studies in the Department of Science and Technology Studies at Rensselaer Polytechnic Institute in Troy, New York, in the United States. His most recent books are *Mathematics in Society and History* (1993), *Science, Society and Values* (1993), and *The Sociological Worldview* (1991).

Michael D. Resnik is University Distinguished Professor of Philosophy at the University of North Carolina in Chapel Hill, in the United States. He has published widely in the philosophy of mathematics and related topics and is the author of *Frege and the Philosophy of Mathematics* (1980).

Ole Skovsmose is associate professor of Education at the University of Aalborg in Denmark, and a member of the BACOMET research group in mathematical education. He is the author of many papers on mathematical education, including "Toward a Philosophy of an Applied Oriented Mathematical Education" (1989), "Mathematical Education and Democracy" (1989), and "Mathematik og Kultur" (1989).

Thomas Tymoczko is professor of Philosophy and Logic at Smith College in Northampton, Massachusetts, in the United States. He received his doctorate in philosophy at Harvard University and has written extensively on the philosophy of mathematics and epistemology. He is currently working on joint projects with his Smith College colleagues in mathematics and computer science.

Subject Index

A

abstraction, 166f., 191, 256ff., 274; and concept of modern mathematics in Nazi Germany, 224; and generational continuity, 254, 258; and iteration, 257, 262ff.; and social organization, 266; Tinker Toy model of, 264ff. *See also* iteration

acting, versus behavior, 142f.

addition, as empirically problematic, 6

aesthetics, 33, 275. *See also* style

algebra, 7, 117, 153, 257

alternative mathematics, in Bloor and Wittgenstein, 252

analysis, classical, 25

Anschauung (geometrical intuition), 228; as metaphor, 229; and Nazi worldview, 229

anti-Semitism, 228, 236ff.

applied systems analysis, 208f.

arithmetic, 72; fundamental theorem of and proofs, 31; modular, 7; ordinary, 7

artificial intelligence, 30f., 176f., 188, 214f.

associative axiom (law), 7

autopoietic systems, 199

Axiom of Choice, 22, 65

axiom of infinity, 95

axioms, 70f., 75, 98, 225f., 263f; Peano-, 130; ZFC-, 26

B

belief, 64, 66f., 72, 189, 192; -community, 188; and knowledge, 172; true, 47

behaviorism, 144

Bishop's constructivist, 25

bureaucratization, 255

C

calculus, 80, 262; Newtonian, 62

capital (cultural); scientific, and productivity, 221

caste system, 252

causality, 43, 46, 114, 214, 226

class consciousness, 124

cognition, 49f., 57, 83f., 89f., 92, 99, 142; a priori, 84; molecular, 90

cognitive science, 49

communication, 49, 57f., 128, 163, 169, 200f., 212, 216, 220, 222, 249

commutative law, 7

complexity, 198

computerization, 114, 118, 121, 173, 183, 188ff., 212, 214

confining inconsistencies, 29

constructivism, 25, 27, 33, 47, 67, 130, 137, 143, 158f., 231; and ethics and politics, 159

Copernican social science revolution, 248

Name Index

A

Aristotle, 6
Armstrong, D., 40ff., 49, 57
Armstrong, J., 134
Austin, J., 14, 170

B

Bamme, A., 120
Baruk, S., 184
Becker, J., 135
Beer, S., 215
Beeson, M., 25
Benacerraf, P., 39, 70, 72f., 75
Berlin, I., 189
Bieberbach, L., 224, 227ff., 239ff.
Bishop, E., 23, 25
Bloor, D., 63, 165, 250f., 268
Blumenfeld, P., 135
Bombelli, R., 263
Bonner, J., 90
Boole, G., 249, 256f., 272
Boos, B., 266
Boswell, S., 135, 137ff., 157
Bourbaki, N., 32, 184, 248, 259
Bourdieu, P., 221
Bowles, S., 274
Bradie, M., 83
Brouwer, L.E.J., 22, 24, 227, 230
Browne, M., 62
Bruffee, K., 188, 192

C

Campbell, D., 12, 82, 84f.
Cantor, G., 67, 70, 74, 77, 80f., 95,
  262
Carnap, R., 129, 169
Carruccio, E., 89
Chipman, S., 134, 157
Church, A., 95
Clark, K., 183
Collins, H., 249
Collins, R., 261
Cronbach, L., 211
Crumb, T., 251
Curry, H., 260

D

D'Ambrosio, U., 162, 275
Darwin, C., 89f.
Davis, P., 3, 6, 14, 63
Dedekind, R., 80, 95, 257
Descartes, R., 64, 66f., 164
Dewey, J., 152, 155
Dobzhansky, T., 100
Dodd, J., 44
Donlon, T., 157
Dostoevsky, F., 4, 8
Douglas, M., 8
Dreyfus, H., 214f.
Dreyfus, S., 214f.
Durkheim, E., 16, 249f., 255f.